Robust Control Algorithms for Two-link Flexible Manipulators

Kshetrimayum Lochan
Research Fellow, Robotics Team
Department of Mechanical and Nuclear Engineering
Khalifa University Center for Robotics and Autonomous Systems (KUCARS)
Khalifa University, United Arab Emirates

Binoy Krishna Roy
Professor and Dean of Academics
Department of Electrical Engineering
National Institute of Technology Silchar (NITS), India

Bidyadhar Subudhi
Director of National Institute of Technology
Warangal (NITW), India

Santhakumar Mohan
Professor, Indian Institute of Technology
Palakkad, IIT PKD, India

CRC Press
Taylor & Francis Group
Boca Raton London New York

CRC Press is an imprint of the
Taylor & Francis Group, an **informa** business
A SCIENCE PUBLISHERS BOOK

First edition published 2025
by CRC Press
2385 NW Executive Center Drive, Suite 320, Boca Raton FL 33431

and by CRC Press
4 Park Square, Milton Park, Abingdon, Oxon, OX14 4RN

CRC Press is an imprint of Taylor & Francis Group, LLC

Library of Congress Cataloging-in-Publication Data (applied for)

ISBN: 978-1-032-38475-7 (hbk)
ISBN: 978-1-032-38478-8 (pbk)
ISBN: 978-1-003-34526-8 (ebk)

DOI: 10.1201/9781003345268

Typeset in Times New Roman
by Prime Publishing Services

For my . . .

Parents and family

Preface

In this book titled "Robust Control Algorithms for Flexible Manipulator", various modelling and control issues of a two-link flexible manipulator are presented. An exhaustive literature survey was carried out. Some salient outcomes of the literature which have motivated us to do this work are presented as follows: (1) Although many controllers are used to control a TLFM, some robust controllers like second order sliding mode control, linear matrix inequality based sliding mode control, adaptive time-varying super-twisting global sliding mode control, have not been applied to control these manipulators. (2) In a two-link flexible manipulator, there exist two dynamics: one characterized by fast dynamics and the other by slow dynamics. The singular perturbation method is employed to distinguish between these two dynamics. While there is existing literature demonstrating the application of singular perturbation for a two-link flexible manipulator, there are opportunities for further exploration and research in this domain. (3) Synchronisation between two manipulators has been extensively documented for rigid manipulators. However, there is a significant scope for work on various types of synchronisation among flexible manipulators. (4) Various desired signals are used for synchronisation of rigid manipulators and a chaotic signal can be considered as the desired trajectory for synchronisation. Considering the title of the book, the scope of work has concentrated on the following.

The lumped parameter method and the assumed modes method modelling of a two-link flexible manipulator are reviewed in detail in this book. The trajectory tracking problem and tip trajectory tracking problem are considered, along with the suppression of tip deflection of the links. An exponentially time varying signal and a chaotic signal are considered as the desired trajectories. The master manipulator is controlled to synchronise with the desired trajectories. The identical/nonidentical (with that of master) slave manipulator(s) are synchronized with the controlled master, allowing the slave manipulator(s) to indirectly follow the desired trajectories. Various synchronisation techniques such as (i) generalised projective synchronisation (complete/scaled) and (ii) projective synchronisation are used in this book. The synchronisation between the controlled master and the slave manipulator(s) is achieved between (i) lumped parameter-lumped parameter modelling, (ii) assumed modes-assumed modes modelling and (iii) assumed modes-lumped parameter modelling of two-link flexible manipulators. Various control techniques such as (i) conventional sliding mode control, (ii) modified adaptive sliding mode control, (iii) adaptive sliding mode control, (iv) linear matrix inequality based sliding mode control, (v) linear matrix inequality based state feedback controller, (vi) backstepping controller, (vii) second order proportional integral derivative terminal sliding mode control, (viii) global sliding mode control and (ix) adaptive time-varying super-twisting global sliding mode control are used to achieve the above-mentioned trajectory tracking and synchronisation. The singular perturbation method is also used to decompose the system into fast and slow subsystems to capture the flexible and rigid dynamics, respectively of the flexible manipulator. All control techniques are used in the presence of variable payloads and parameters uncertainty to check the robustness of the proposed controllers. The results obtained from the proposed controllers are compared with those of existing controllers, demonstrating superior performance.

Contents

List of Figures

List of Tables

Abbreviations

AASFC	Adaptive Augmented State Feedback Control
AE-SMC	Adaptive Equivalent Sliding Mode Control
AMM	Assumed Modes Method
ANN	Artificial Neural Network
ARMAX	Auto Regressive Moving Average Exogenous
A-SMC	Adaptive Sliding Mode Control
A-ST-GSMC	Adaptive Super-Twisting Global Sliding Mode Control
ATDNN	Adaptive Time Delay Neural Network
A-TV-ST-GSMC	Adaptive Time Varying Super-Twisting Global Sliding Mode Control
DAC	Direct Adaptive Control
DOF	Degrees Of Freedom
FLAC	Fuzzy Learning and direct Adaptive Control
FLC	Fuzzy Logic Control
FM	Flexible Manipulator
FSOSMC	Frequency Shaped Optimal Sliding Mode Control
GA	Genetic Algorithm
GPS	Generalised Projective Synchronisation
GSMC	Global Sliding Mode Control
HONTSMC	Higher Order Nonsingular Terminal Sliding Mode Control
HOSMC	Higher Order Sliding Mode Control
LMI	Linear Matrix Inequality
LPM	Lumped Parameter Method
LQG	Linear Quadratic Gaussian
LQR	Linear Quadratic Regulator
MAE-SMC	Modified Adaptive Equivalent Sliding Mode Control
MIMO	Multi Input Multi Output
NAMPC	Nonlinear Adaptive Model Predictive Control
NN-MPC	Neural Network based Model Predictive Control

NSMC	Normal Sliding Mode Control
PD	Proportional Derivative
PDE	Partial Differential Equation
PID	Proportional Integral Derivative
PI-SMC	Proportional Integral Sliding Mode Control
QFT	Quantitative Feedback Theory
RL	Reinforcement Learning
RLS	Recursive Least Square
SFC	State Feedback Control
SMC	Sliding Mode Control
SO-PID-TSMC	Second Order Proportional Integral Derivative Terminal Sliding Mode Control
SO-SMC	Second Order Sliding Mode Control
SPR	Strictly Positive Real
STC	Self Tunning Controller
STR	Strong Tracking Filter
TD	Temporal Difference
TLFM	Two-Link Flexible Manipulator
TSMC	Terminal Sliding Mode Control
VSC	Variable Structure Control

Chapter 1

Introduction

This chapter introduces the flexible manipulator, its challenges and applications. It discusses the foundation of this book and gives a reflection of the content. The chapter is organised as follows. The introductory section provides background information on a two-link flexible manipulator (TLFM) depicting its advantages over rigid manipulators, outlining the associated control complexities, and presenting its benefits. The next section offers the motivation for the present work, outlines the objectives, and concludes with an organisation of the book.

1.1 Background

A robotic manipulator consists of a series of links and joints with different combinations. In most of the industrial applications, rigid links are used along with the joints, forming what is commonly referred to as a rigid manipulator. Rigid manipulators are made up of strong and bulky materials and hence, they are heavy in general. Rigid manipulators are extremely rigid, massive, sluggish in dynamics, low useful load to weight ratio and are generally slow. To improve the performance of robot manipulators, the drawbacks of a rigid manipulator are eliminated by making the links light. Such a manipulator whose links are constructed with thin and light arms is called as a flexible-link manipulator or flexible manipulator (FM). Depending upon the number of links, it is called as a single-link, two-link or multi-link flexible manipulator. Since a flexible manipulators is light in weight, it illustrates many advantages such as higher payload to weight of links ratio, a large increase in the speed of the link, etc. It also requires less power to produce the same acceleration, smaller and cheaper actuators as compared with its rigid counterpart. Flexible manipulators are used in the areas where rigid ones may not be suitable such as pick and place operation of an industrial robot [1], microsurgery [2], submarine [3], aerospace [4], education [5] and maintenance of nuclear plants.

In recent years, researchers over the world have been interested in the technical advancement and development of flexible manipulators as these are motivating and interesting. Research in this field is on going, driven by significant advancements in automation and robotics, which are primarily driven by industrial needs such as heightened productivity, reduced energy consumption, quicker response times and improved accuracy. Flexible manipulators have potential applications in space technologies and exploration. Robots play a major role in space missions which carry out planetary explorations with unmanned space probes. These missions encompass a wide array of tasks including, orbital constructions of space stations, servicing space stations during their operation, collection of the space station during operation, collecting space debris, retreiving spacecraft stranded in orbit, launching satellites, and facilitating cargo movement, among other objectives. Hence, to expand the workspace, reduce fuel consumption, and prolong operational life, the manipulators should be light weight in structure with long arms that introduce link flexibilities [6]. These flexibilities may cause vibrations in the manipulator as well as the spacecraft posing difficulties in positioning the tip of the manipulator. These challenges motivate the development of control techniques for flexible manipulators.

Controlling a flexible manipulator is challenging due to the fact that the link flexibility is distributed which introduces complexities to the system. Controlling a flexible manipulator is much more difficult as compared with a rigid one. The control complexities for a flexible manipulator are as follows:

1. Non-minimum phase

Non-minimum phase is a direct result of the system zeros in the right half of the s-plane. The following points explain the cause of non-minimum phase in a FM:

* Non-collocated sensing and actuation. For instance, the transfer function between the torque input to the tip position output of a flexible manipulator is non-minimum phase [7].

* One input/state map of flexible manipulators is not externally feedback linearisable and the input torque (voltage) at the joint exhibits a non-minimum phase property [8].

The effects of non-minimum phase in a FM are as follows:

* With a bounded control input, it affects the asymptotic tracking of a desired tip trajectory.

* Large tracking error.

* Complicates the control synthesis and limits the achievable bandwidths [9].

2. Underactuation

The term underactuation in case of a robot, refers to the manipulator which has less actuators than the number of joints or (Degrees of freedom) DOF. Some common examples of underactuation robot systems are: (a) legged and snake like robots with passive joints, (b) free floating space robots, etc. The importance and common properties of an underactuation system are:

* Dynamic coupling: According to [10], DOF of a flexible manipulator are coupled with dynamic forces and control of such system requires the existence of dynamic coupling.

* The controllability of the unactuated DOF [11]: All DOF of such systems can be controlled indirectly or directly by actuated DOF.

* Nonholonomic constraints: In a nonholonomic system, generalised coordinates do not depend upon their time derivative as seen in [12].

* Nonlinearity and dependency on dynamic parameters: Dynamic equations are nonlinear and depend on the structure and inertial characteristics of the system as reported in [13].

The causes of underactuation are as follows:

* Failure of actuators - An example is a conventional robot with failed actuators.

* Functionality requirement - Examples are a hyper-redundant snake like robot and a human body.

* Inherent property of the system - examples are free floating space satellites and flexible manipulators.

The effects of underactuation on a FM are as follows:

* Energy saving.

* Reduction in weight and improved compactness.

* Requires enhanced control techniques.

In case of flexible manipulators, most of the research focuses on deals with underactuation systems. However, some studies have explored various techniques to change the characteristics of the dynamics of the system [14].

3. Non-collocation

The noticeable problem with a flexible manipulator is controlling the end-point of the system accurately [15]. The main reasons for the non-collocation in FMs are as follows [16]:

* Distinct location of sensors and actuators.

* Pole-zero alteration.

The effects of non-collocation in a FM [16]:

* Non-minimum phase in the system.
* Critical accurate dynamic modelling.
* Accurate modelling of zeroes is critical near the system bandwidth.
* In a non-collocated system, model inaccuracies and model truncation are present, which lead to inaccurate system performance and system stability.

In 1990, the authors of [16] did experiments on the sensitivity effects of the structural models for non-collocated control systems. It is shown that the accurate dynamic modelling of a non-collocated control system is critical near the system bandwidth. According to [16], some finite frequency (this frequency is dictated by the location of the nearest right hand side (RHS) zero, all non-collocated systems are non-minimum phase and as the distance between the sensor and actuator increases, this frequency also increases. The hint for an accurate numerical solution of a non-collocated control problem dealing with the RHS zeros is to have a good system model. Often RHS zeros are called unstable zeros, but they do not make the system unstable. RHS poles cause the system response to increase exponentially resulting in instability. But, the system zero will have an effect on the system stability depending on its controller design [17]. In the case of a non-minimum phase system, its possible for the feedback controller to ensure perfect output tracking of the reference signal. However, this achievement in output tracking might coincide with system states being unstable. The instability of the system is not reflected in the output, which is dangerous [17]. The presence of a RHS zero imposes a maximum bandwidth limitation. It means the bandwidth of the system cannot be more than the absolute value of the zero. If we move the bandwidth frequency close to the zero, it gives a very high sensitivity function peak, meaning that disturbance rejection of the system is limited.

The uncertainties of flexible manipulators are as follows:

1. Truncation of flexible modes.

Model truncation which arises due to a finite dimensional representation of a distributed parameter system, causes unmodeled dynamics to be present in the mathematical model of the system. Using a reduced order model, the controller design may also lead to the phenomena of control and observation spillover. Control spillover is the excitation of the residual modes by the controller action and observation spillover is the contamination of the sensor reading by the residual modes. The closed loop system may become unstable when the observation and control spillover are present. The rigid and flexible dynamics coupling also causes a stability problem.

2. MIMO system

MIMO (multiple inputs multiple outputs) concept is used in terms of the robot manipulator when the system has several flexible links in its structure. Such a system is considered to be highly non-linear with a distributed parameters control problem [18]. The following are the reasons of a MIMO structure in FMs:

* Require more DOFs.
* Accuracy in modelling.
* Nonlinear distributed structure.

The effects of the MIMO structure in a FM are as follows:

* Easy operation handling.
* Require a precise modelling method.
* Control complexities increase.

Considering the control problem and uncertainties described above, it is seen that the control of a flexible manipulator is challenging due to the oscillations involved in the distributed flexibility across the entire link. Hence, owing to the oscillations in the link, achieving accuracy in the positioning of the tip deflections is difficult. Therefore, in order to improve the tip position or tip trajectory tracking of a flexible manipulator, elastic properties have to be taken into account while designing the control techniques. Furthermore, as the manipulator is expected to track the desired path with unanticipated payloads, their variability at the end of the link is an important concern. Therefore, due to this sudden change in the payload, there may be a large variation in the parametric values of a controller which further complicates the dynamics as well as the control of the flexible manipulator. Thus, the control techniques are to be designed such that the effect of the change in the payload is taken into account.

1.2 Motivation of the present book

Motivation of this work has come from the outcome of the literature survey which is presented in the next chapter. The literature survey has reflected the research gap in the area of modelling, control and synchronisation of a two-link flexible manipulator. Most of the control techniques on TLFMs use the assumed modes method for designing the controllers. But, the designing of a controller for a two-link flexible manipulator with singular perturbation using the assumed modes method is found to be predominately less explored. Hence, there is an immediate motivation to design the manipulator dynamics using the singular perturbation method.

It is also seen that among the mentioned controllers in the literature, sliding mode control gives better results due it its advantages like robustness to uncertainties, insensitivity to disturbances, global stability and convergence of system dynamics. Normally, in conventional sliding mode control, chattering, high gain and asymptotic stability [19] are considered as disadvantages. Nevertheless, for first order SMC, chattering can be reduced by using a saturation function instead of a signum function. But, it is seen that the precision tracking and disturbance rejection properties get degraded due to the bounded sliding mode variables [19]. An integral SMC is proposed in [20] to overcome the high gain requirement. The terminal sliding surfaces are designed to achieve finite time stability [19]. Terminal sliding mode control (TSMC) techniques are also used for different control problems of a TLFM [21]. However, if the initial states of the system are distant from the equilibrium point, a good convergence performance may not be achieved by a TSMC. Another disadvantage of the first order TSMC is the singularity problem which requires a large control effort [22]. Many methods are available to reduce or to eliminate the chattering like quasi-SMC, low pass filters, Fuzzy-SMC and higher order SMC (HOSMC) [22]. Thus, there is a motivation to propose a second order sliding mode controller owing to the elimination of chattering as the actual control is a continuous intergration of its derivative.

Moreover, in the literature, different types of desired trajectories such as bang-bang trajectory [23], circular trajectory [24], sinusoidal trajectory [25], exponentially varying signal [26], straight line trajectory [27] are used. This motivates us to use a chaotic signal, a seemingly random-like signal, as a desired trajectory for a two-link flexible manipulator.

For the purpose of an application, the effective operation of serial manipulators requires an accurate and coordinated control algorithm. Coordination, cooperation and synchronisation are synonym used to define the mutual temporal alignment and cooperation between two processes [28]. However, some of the inherent disadvantages [29] of rigid manipulators limit the efficient applications of the serial flexible manipulators. Thus, serial manipulators with coordinated flexible links are more desirable. A significant amount of research has been focused in the past on the synchronisation of rigid robot manipulators [20]. The synchronisation of robot manipulators is categorically explained in [28], [30]. It is notable that most of the available synchronisation concepts are focused on the rigid manipulators which are generally not precise, economical and effective in industrial applications. There is no paper on the synchronisation of two-link flexible manipulators. So, there is an opportunity in designing of an appropriate control technique for the synchronisation of two-link flexible manipulators.

1.3 Objectives of the book

The objectives of this book are as follows:

* To derive the dynamic models of a TLFM and to validate the theoretical model to be used for the proposed control techniques.

* To design and implement both first-order and second-order sliding mode controllers for effectively controlling the tip trajectory tracking of a TLFM. This control mechanism should be robust against varying payloads the TLFM may carry, while also effectively suppressing any deflection in its structure.

* To explore the model decomposition of a TLFM by the singular perturbation method and to design the controllers of the slow and the fast subsystems subjected to varying payloads. Trajectory tracking is performed by the slow subsystem while tip deflection suppression is done by fast subsystem. The performance of the designed controller is then compared with an existing controller.

* To design the modified adaptive equivalent SMC for a generalised projective synchronisation between lumped parameter model TLFMs for a good trajectory tracking performance along with quick suppression of tip deflection in the presence of varying payloads and compare its performance with other controllers.

* To design a new second-order PID terminal sliding mode controller for synchronisation between assumed modes model TLFMs and study the performance of tip trajectory tracking and its deflection in the presence of parameter uncertainties and varying payloads at the end-effector.

* To design the projective synchronisation between the controlled assumed modes modelled master TLFM and three lumped parameter modelled slave TLFMs (two identical and one non-identical with the master) for controlling the tip trajectory and tip deflection suppression in the presence of variable payloads.

1.4 Organisation of the book

Chapter 1 Gives a background of a flexible manipulator highlighting its advantages over rigid manipulators. The control complexities associated with the flexible manipulator are presented. Motivation, objectives and organisation of the book are also described in this chapter.

Chapter 2 Gives an overview of flexible manipulators. It includes a literature survey on the review of various control techniques of TLFMs and perturbation based controllers highlighting their applications.

Chapter 3 Develops two types of mathematical models of a TLFM. The resulting theoretical mathematical model is validated both in free and forced conditions. The nonlinear dynamic equations of the TLFM are used for the development of different controllers and synchronisation techniques in the subsequent chapters.

Chapter 4 Develops a conventional sliding mode controller (SMC) and a second order SMC. The results using both the controllers reveal that the second order SMC outperforms the conventional SMC in terms of tip trajectory tracking errors and tip deflection suppression in the presence of varying payloads.

Chapter 5 Describes model decomposition by singular perturbation. The nonlinear model is based on the assumed modes method. Subsequently, the controllers of the fast and the slow subsystems are designed for the developed mathematical model in the presence of payloads. The results show that the trajectory tracking errors and tip deflections by the designed controllers are minimised as compared to the existing controller in the literature.

Chapter 6 Provides a generalised projective synchronisation between the lumped parameter modelled TLFMs. A new modified adaptive equivalent SMC is designed with a scaling factors of 1, 0.5, 0.25 for the synchronisation technique between the controlled master and the slave manipulators. The results show that the proposed technique has faster trajectory tracking along with the suppression of tip deflection with varying payloads as compared with the other three controllers, i.e. (i) conventional SMC with tanh($\frac{s}{\epsilon}$), (ii) adaptive equivalent SMC with tanh ($\frac{s}{\epsilon}$) and (iii) modified adaptive equivalent SMC with sigmoid ($\frac{s}{|s|+\epsilon}$).

Chapter 7 Develops a new second order PID terminal sliding mode controller for the synchronisation of TLFMs. A conventional SMC is designed for the controlled master. Parametric uncertainties of $\pm 5\%$ and $\pm 30\%$ are induced to check the robustness of the designed controller. The results show a significant reduction in the tip trajectory tracking errors and minimising the magnitude of tip deflections when the system is subjected to varying payloads are as compared with a standard/normal second order SMC.

Chapter 8 Develops a projective synchronisation between the controlled assumed modes method modelled master TLFM and three lumped parameter modelled slave TLFMs (two of them are identical and one non-identical with the master manipulator). The master manipulator having twelve states are projected to slaves having eight states for the synchronisation. Subsequently, an adaptive time-varying super-twisting global SMC is designed for the projective synchronisation. The results show that the tip trajectory tracking and tip deflection suppression are minimised when compared with an existing controller in the literature.

Chapter 9 Summarises the work described in this book and the contributions. This chapter also provides a brief overview of potential areas for further research that could be pursued in future building upon the themes covered in this book.

References

[1] C. T. Kiang, A. Spowage, and C. K. Yoong. Review of Control and Sensor System of Flexible Manipulator. *J. Intell. Robot. Syst. Theory Appl.*, vol. 77, no. 1, pp. 187–213, 2015.

[2] R. J. Hendrick, C. R. Mitchell, S. D. Herrell, and R. J. Webster. Hand-held Transendoscopic Robotic Manipulators: A Transurethral Laser Prostate Surgery Case Study. *Int. J. Rob. Res.*, vol. 34, no. 13, pp. 1559–1572, 2015.

[3] R. Rout and B. Subudhi. Inverse Optimal Self-tunning PID Control Design for an Autonomous Underwater Vehicle. *Int. J. Syst. Sci.*, vol. 48, no. 2, pp. 367–375, 2017.

[4] T. Rybus, K. Seweryn, and J. Z. Sasiadek. Control System for Free-Floating Space Manipulator Based on Nonlinear Model Predictive Control (NMPC). *J. Intell. Robot. Syst.*, vol. 85, no. 3, pp. 491–509, 2017.

[5] B. Subudhi and S. K. Pradhan. A Flexible Robotic Control Experiment for Teaching Nonlinear Adaptive Control. *Int. J. Electr. Eng. Educ.*, vol. 53, no. 4, pp. 341–356, 2016.

[6] H. Geniele, R. V. Patel, and K. Khorasan. End-point Control of a Flexible-link Manipulator: An Experimental Study. *IEEE Trans. Contr. Syst. Technol.*, vol. 5, pp. 556–570, 1997.

[7] D. Chen and B. Paden. Stable Inversion of Nonlinear Non-minimum Phase Systems. *International Journal of Control*, vol. 64, pp. 81–97, May 1996.

[8] N. Zhang, Y. Feng, and X. Yu. Optimization of Terminal Sliding Control for Two-link Flexible Manipulators. In 30th *IEEE annual Conference of the Industrial Electronics Society*, vol. 2, (Busan, Korea), pp. 1318–1322, November 2004.

[9] F. Khorrami and S. Jain. Experimental Results on an Inner/Outer Loop Controller for a Two-Link Flexible Manipulator. In *Proc. of the IEEE International Conference on Robotics and Automation*, vol. 1, pp. 742–747, May 1992.

[10] X. Bo, Y. Fujimoto, and Y. Hayakawa. Control of Two-link Flexible Manipulators via Generalized Canonical Transformation. In *IEEE Conference on Robotics Automation and Mechatronics*, vol. 1, pp. 107–112, December 2004.

[11] M. Bergeman. *Dynamic and Control of Underactuated Manipulators.* PhD Thesis, Carnegie Mellon University, 1996.

[12] W. Chen, Y. Yu, L. Zhao, and Q. Sun. Position Control of a 2DOF Underactuated Planar Flexible Manipulator. In *International Conference on Mechatronics and Automation (ICMA)*, (Beijing), pp. 464 – 469, August 2011.

[13] X. Xin. Swing-up Control for a Two-Link Underactuated Robot with a Flexible Elbow Joint: New Results beyond the Passive Elbow Joint. In 50th *IEEE Conference on Decision and Control and European Control Conference (CDC-ECC)*, (Orlando, FL, USA), pp. 2481 – 2486, December 2011.

[14] X. Xin, Y. Liu, and J. Wu. Global Stabilization Control for a Two-Link Underactuated Robot with a Flexible Elbow Joint. In 32nd *Chinese Control Conference (CCC)*, (Xian), pp. 1520 – 1525, July 2013.

[15] H. A. Talebi, R. V. Patel, and K. Khorasami. *Control of Flexible-link Manipulators Using Neural Networks.* Springer, 2001.

[16] V. A. Spector and H. Flashner. Modeling and Design Implications of Noncollocated Control in Flexible Systems. *Journal of Dynamical System Measurement and Control*, vol. 112, pp. 186–193, June 1990.

[17] M. Karkoub, G. J. Balas, and K. Tamma. Colocated and Noncolocated Control Design via μ-Synthesis for Flexible Manipulators. In *Proc. of the American Control Conference*, vol. 5, (Seattle, WA), pp. 3321 – 3325, June 1995.

[18] F. Wang and Y. Gao. Advanced Studies of Flexible Robotic Manipulators Modeling, Design, Control and Applications. In *Series in Intelligent Control and Intelligent Automation*, vol. 4, 2003.

[19] S. Ding, J. Wang, and W. X. Zheng. Second-Order sliding mode control for nonlinear uncertain systems bounded by positive functions. *IEEE Transactions on Industrial Electronics*, vol. 62, no. 9, pp. 5899–5909, 2015.

[20] D. Zhao and Q. Zhu. Position synchronised control of multiple robotic manipulators based on integral sliding mode. *International Journal of Systems Science*, vol. 45, no. January 2014, pp. 556–570, 2014.

[21] Y. Wang, C. Wang, P. Lu, and Y. Wang. High-order nonsingular terminal sliding mode optimal control of two-link flexible manipulators. *IECON Proceedings (Industrial Electronics Conference)*, pp. 3953–3958, 2011.

[22] X. T. Tran and H. J. Kang. Adaptive hybrid high-Order terminal sliding mode control of MIMO uncertain nonlinear systems and its application to robot manipulators. *International Journal of Precision Engineering and Manufacturing*, vol. 16, no. 2, pp. 255–266, 2015.

[23] A. Sanz and V. Etxebarria. Experimental control of a two-dof flexible robot manipulator by optimal and sliding methods. *Intell Robot Syst*, vol. 46, pp. 95–110, 2006.

[24] L. Zhang and J. Liui. Observer-based partial differential equation boundary control for a flexible two-link manipulator in task space. *IET Control Theory Appl.*, vol. 6, pp. 2120–2133, 2012.

[25] W. Yu, M. Karkoub, T. Wu, and M. Her. Delayed output feedback control for nonlinear systems with two-layer interval fuzzy observers. *IEEE Transactions on Fuzzy Systems*, vol. 22, pp. 611–630, 2014.

[26] S. Pradhan and B. Subudhi. Real-time adaptive control of a flexible manipulator using reinforcement learning. *IEEE Transactions on Automation Science and Engineering*, vol. 9, pp. 273–249, 2012.

[27] Y. Li, G. Liu, T. Hong, and K. Liu. Robust control of a two-link flexible manipulator with quasi-static deflection compensation using neural networks. *Journal of Intelligent and Robotic Systems*, vol. 44, pp. 263–276, 2005.

[28] A. Rodriguez-Angeles and H. Nijmeijer. Mutual synchronization of robots via estimated state feedback: A cooperative approach. *IEEE Transactions on Control Systems Technology*, vol. 12, no. 4, pp. 542–554, 2004.

[29] B. Subudhi and A. Morris. Soft computing methods applied to the control of a flexible robot manipulator. *Applied Soft Computing*, vol. 9, no. 1, pp. 149–158, 2009.

[30] D. Zhao, S. Li, and Q. Zhu. Adaptive synchronised tracking control for multiple robotic manipulators with uncertain kinematics and dynamics. *International Journal of Systems Science*, vol. 7721, no. May 2014, pp. 1–14, 2014.

Chapter 2

Survey on a Two-link Flexible Manipulator

This chapter presents a survey on the modelling of TLFM and explores various synchronisation and control techniques proposed for TLFM. Due to the novel robotic applications, the demand for the lighter robotic manipulators driven by the small energy requirement is increasing in aerospace industries where high speed operation, light weight, better accuracy, high payload to weight ratio is required [1]. But, due to its light weight, flexibility in the links of the manipulator leads to oscillatory behaviour at the tip of the manipulator making tip positioning or precise pointing problematic requiring complex closed loop control. The structure of an flexible manipulator (FM) is also an important factor that needs to be focused while discussing the complexities associated with the dynamic model. Generally, the structure of an FM depends on its required operations. The operation of FMs is basically its control. The major categories of the control problems for flexible manipulators can be classified as: tip position control [2], joint position control [3], tip trajectory tracking control [4], joint trajectory tracking control [5], vibration control [6], motion control [7], force control [8], hybrid control problem, etc. Different control techniques are developed depending upon the type of the control problems like PID control [5], feedback control [9], LQR control [10], observer based control [11], SMC/VSC [12], adaptive control [13], H_∞ control [14], backstepping control, optimal control [10], LQR control, fuzzy logic control (FLC) [15], artificial neural network (ANN) based control [16], GA based control technique [17], QFT control [18], hybrid control techniques, etc.

2.1 Review on various control techniques of a TLFM

This section provides a comprehensive literature survey on various control methods reported for addressing control issues related to Two-Link Flexible Manipulators (TLFM).

1. Classical control techniques

PID control, state feedback control (positive position feedback, negative position feedback), LQR (Linear Quadratic Regulator) and observer based control are generally considered as classical control techniques.

PID control: PID control is a non model based control technique. Simplicity, reliability and broad applicability are the main advantages of this control. The PID control technique is given in (2.1).

$$u = k_p e + k_i \int edt + k_d \frac{de}{dt} \tag{2.1}$$

where e is the error defined and u is the control input. Reference [5] presented the AMM modelling and trajectory tracking control of a TLFM using PID controller. Reference [19] reported the hybrid collocated PD and noncollocated PID control for input tracking control of a TLFM. Lagrange's and AMM are used for modelling of a two-link flexible manipulator (TLFM). The performances of the reported control technique are also shown using different payloads. The performances are also evaluated in terms of input tracking capability, energy utilization and vibration suppression.

The position control of an underactuated TLFM in joint space is presented in [3]. AMM is used for modelling the dynamics. Results are also verified using an experimental setup. Modeling and control of a telescoping TLFM are presented in [20]. Telescoping interference between the deploying and non deploying portions of the links is discussed, and tip position control of the proposed manipulator is achieved by using a PD type controller.

The coupled dynamics of a two-link piezoelectric flexible manipulator is developed in [21] using AMM and Lagrange equations with the help of feedback control. The stress on the surface layer of the second link is obtained through a modal function for optimal placement of the actuators on the second link.

LMI based PD control: Reference [22] reported the position control and tip deflection suppression for a TLFM using a LMI based PD controller. The robustness of the proposed control technique is checked under varying payloads.

Computed torque method: The computed torque method (extension of feedback linearisation) is used in [23] for position control of a TLFM. Numerical simulation of the proposed method is presented in the presence of uncertainty and disturbances. Reference [24] reported the tip position control of a TLFM using computed torque technique and feedback control. The dynamic model is derived using AMM. Position and tip position control of a TLFM is reported in [25] using the computed torque control technique. AMM is used to derive the dynamics of the manipulator.

State feedback control (positive position feedback, negative position feedback): Generally, a feedback controller is defined as

$$u = kq \tag{2.2}$$

where k is the position feedback gain and q is the position.

Control of a two-pinned beam flexible manipulator is presented in [26]. Three types of linear feedback control schemes are considered. Joint angle and velocity feedback with (general rigid control), without (independent joint control) cross joint feedback, and feedback of the state variable (flexible feedback control) are the control techniques used.

In [2], a state feedback control is used for tip position control and vibration suppression of a TLFM. The proposed controller is applied separately and in composite form for the slow and fast subsystems. Dynamics are divided into slow and fast subsystems using singular perturbation techniques. Dynamics analysis of a TLFM using Finite Element Method (FEM) is reported in [27]. The controller is designed for rigid body motion using pole placement techniques. Different payloads are also considered for model validation. Reference [9] reported a robust state feedback control for trajectory tracking control of a TLFM flexible pendulum with two degrees of freedom. The attractive ellipsoid method is designed using nonlinear bounded feedback for "quasi-Lipschitz" dynamics of a FM. The feedback parameters are optimized using the LMI toolbox. Tip trajectory tracking control of a TLFM using feedback control is presented in [28]. The manipulator model is derived using the AMM with causal stable inverse dynamics. In [29], the position and vibration control of a TLFM using state feedback control technique is reported. The dynamic model of the flexible manipulator is derived using the AMM. The inverse dynamic motion equation of a planar TLFM is discussed in [30]. A computed torque control is used for trajectory tracking control of a TLFM.

The trajectory tracking control of an AMM modeled TLFM using feed forward control scheme is considered in [31]. The control law consists of two parts, commanded feedforward torque and PD controller. The proposed control technique in [6] is designed for rigid body motion. The lumped parameter method is used for modelling the dynamics. In [32], force/position and vibration control of a TLFM are proposed using PD and feedback control. The model is separated into lumped parameters and distributed parameters. A vision sensor based motion control of a TLFM is presented in [33]. Feedback signals are also generated using a vision sensor and a PID controller is used to control the motion of a TLFM. Reference [34] reported an approach which is actively involved in suppressing the vibration within a two-link flexible manipulator to adapt the variation in the model parameters, which are composed of an input shaper and multimode adaptive positive position feedback. The input shaper is applied to shape the command to avoid the flexible vibration in the manoeuvre motion and the residual vibration is suppressed by a piezo actuator with the adaptive positive position feedback approach.

Optimal linear quadratic regulator, (LQR) control: This is a model based control technique. The optimal performances are the main advantages of this technique. The general design structure of this control technique is given in (2.3) to minimize the cost function and to find a gain matrix given by (2.4).

$$J = \int_0^\infty (x^T Q x + u^T R u) dt \tag{2.3}$$

$$u = -k_c x \tag{2.4}$$

where k_c is the gain matrix, Q and R are the positive definite matrices. The trajectory tracking control for a TLFM is presented in [10] using a minimum time control solution with actuator constants. An optimal control is also used in combination with other classical controls for different control problems of a TLFM. The tip position control of a TLFM is achieved in [35] using tip position acceleration feedback. The control technique used in [35] consists of two stages. A feedback linearisation controller is designed for rigid body motion and a LQR is designed for the flexible part of the dynamics. Comparison of the proposed controller with feedback linearization and PD controller is also presented. The problem of achieving rest to rest motion within a fixed desired time for a TLFM is described in [28]. A Feedforward and state feedback control is used for tip position control and vibration suppression of the manipulator. The drawback of the state feedback controller is overcome by designing a LQR controller.

Observer based control: This is a non model based control technique. The important advantage of this observer based control technique is that a state feedback control technique can be used even when some of the states are not measurable. The structure of an observer based control technique is given in (2.5).

$$\dot{\hat{x}} = A\hat{x} + Bu + LC(x - \hat{x}) \tag{2.5}$$

where A and B are the original system matrix and input matrix, respectively. Here, u is the control input and L, $\hat{x}$ are the observer gain matrix and estimated state vector, respectively. C is the system output matrix. The tip position tracking control for a TLFM is presented in [11] using a nonlinear observer. The Lyapunov stability theory is used for the stability analysis. The AMM is used for modelling the dynamics of the system. The trajectory control of a TLFM is reported by [36] using a partial differential equation (PDE) dynamic model. The nonlinear PDE observer is used to estimate the position and links of the manipulator. A boundary control scheme is used for tip trajectory tracking and vibration suppression of the FM. An observer is designed for a TLFM using a PDE dynamic model in [37]. The proposed observer can estimate the infinite dimension along the links and is used for position control of a TLFM.

2. Control approach

The control approach can be classified into model-based and non-model-based controllers. In the model based approach, the structure and the vibrational effect are considered carefully to suppress the vibration. These controllers are usually feed-forward and no feedback sensors are required. The main drawback is that it does not account for the changes in the system. The model based controllers include: input shaping or command shaping, optimal control, optimal trajectory planning, boundary control and predictive control.

The non-model based controllers are easier to implement and are usually based on feedback control. The estimation of the states are taken from the sensors to regulate the control input in order to minimise the system vibrations. The main drawback is the input delay in the feedback loop. The non-model based controllers include linear velocity feedback, PID, direct strain feedback, fractional order, singular perturbations, sliding mode control, robust, adaptive, neural networks and fuzzy logic.

3. Robust control

Variable structure control, sliding mode control (SMC), adaptive control, H_∞ control, backstepping control and disturbance observer are the main types of robust control techniques. In real practice, the dynamics of the plant are affected by the uncertainties (such as model uncertainty and parametric uncertainty) and disturbances (associated with input, output and plant). Thus, it requires robust control techniques to deal with these extra challenges associated with the plant.

Sliding mode control (SMC) and variable structure control (VSC): It is a non model based robust control technique. It is robust to model uncertainty, inaccuracy, consistency in performance, order reduction, and insensitive to parameter variation. The design of SMC consists of two steps: the first step is to design a sliding surface/sliding variable and the second is to design a sliding mode control law. Various structures of the sliding variable are proposed in the literature. Generally, the sliding variable is defined in the form as given in (2.6) and (2.7).

$$s = \dot{q} + cq \tag{2.6}$$

or,

$$\sigma = \dot{q} + cq \tag{2.7}$$

s and σ are the sliding surfaces, q is the position vector and c is a constant specified by the designer. The structure of a sliding mode control law is,

$$u = u_{eq} + u_c \tag{2.8}$$

where u_{eq} is the equivalent control input and u_c is the corrective control input. A proper design of the SMC is given in Chapter 4.

A variable structure control is used in [38] for joint angle trajectory control of a TLFM. Robust tracking control of a TLFM is achieved in the presence of payload uncertainty. A terminal SMC is designed in [39] for tip position control using the output redefinition method for a TLFM. In this, the manipulator dynamics are divided into input-output dynamics and zero dynamics. SMC is also used in combination with a PID control for different control problems of a TLFM. Trajectory tracking control of a TLFM is presented in [40] using the conventional SMC. A distributed control strategy is designed in [41] for trajectory tracking and vibration control of a TLFM using numerical methods. The distributed control strategy is designed in [41] decomposing the dynamics into two subsystems. Each subsystem consists of one-link and one joint. Conventional SMC is used to control each subsystem. The results are simulated and compared with the results of the PID controller. Tip position control of a TLFM is given in [42] by using a hybrid actuator scheme. The AMM is used for modelling the dynamics of the manipulator. The proposed control scheme in [42] combines the actuators using two servo motors at the hub and two piezoceramics attached on the surface of the link. Two sliding hyperplanes with time varying parameters are designed for the two servo motors. The pole placement technique is used to design the hyperplane variable. A feedback control voltage is applied on the piezo ceramic actuators to suppress the undesirable torque caused by the rigid link dynamics. Trajectory tracking control of a TLFM is proposed in [12] using an optimal (LQR) and SMC control. The dynamics of the manipulator are divided into slow and fast subsystems. LQR and SMC are used to design different control strategies for slow and fast subsystems, respectively.

Adaptive control: Direct and indirect adaptive controls are the main types of adaptive control. Insensitive to parametric uncertainties, varying payload or unknown disturbances are the main advantages of an adaptive control. This parameter adaptation mechanism is estimated directly from the motion tracking error in the case of direct adaptive control. On the contrary, in indirect adaptive control, parameters of the system are updated using online estimation. The structure of the direct adaptive control for a two-link flexible manipulator is given as under in [13].

$$\tau_i = [Y(q, \dot{q}, \dot{q_r}, \ddot{q_r})]^T \hat{a} - K_D s_i \tag{2.9}$$

where τ_i is the applied torque to the flexible manipulator, $\hat{a}$ is the parameter estimation error, K_D is the positive definite matrix, s_i is the measure of tracking accuracy of the link, q is the tip position and q_r is the reference velocity [13].

Decentralized gain scheduling adaptive control is used by [43] for tip position control of a TLFM. The control scheme is also applied for handling unknown payload and vibration suppression. Tip position control of a TLFM is discussed in [44] using self-tuning adaptive control with unknown payload. System identification is also presented. Reference [45] reported the position control of a TLFM using adaptive control based on a PDE model. It is shown that neglecting the modes can cause a spillover problem. The controller effectiveness also shows a desirable vibration suppression. Adaptive boundary control in the presence of parametric uncertainties is proposed in [45] for position control and vibration suppression of a TLFM. Position control of a TLFM is presented in [46] using adaptive control. The adaptation law is designed using the smooth projection algorithm and dynamic certainty equivalence principle (computed torque method). AMM method is used to model the dynamics of the manipulator. Adaptive augmented state feedback control is designed in [47] for tip trajectory tracking control of a TLFM. The controller is designed using LQR and an adaptive compensator. An adaptive compensator is designed using a strong tracking filter for estimation of the states. The identification and control of a very flexible TLFM is presented in [48]. A parametric frequency domain identification technique is applied to obtain the linear SISO model. A gain scheduled compensator, i.e., LQG is developed on the linearized model to control the joint angle and the results are validated using numerical simulations and experiments. Position control of a PDE modeled TLFM is presented in [49] using the adaptive boundary control technique. The PDE model is developed to reduce the spillover problem caused by neglecting flexible modes. Vibrations are also suppressed using the proposed technique. Model reference adaptive control is designed for the position and trajectory tracking control of a TLFM [50]. A feedforward compensator is designed for validating the linear model of nonlinear dynamics. AMM is used for modelling the dynamics of the manipulator. Two types of adaptive energy based controllers are proposed by [51] for position control of a TLFM. The first controller consists of a joint PD controller and an adaptive proportional controller to suppress the vibration. The second controller consists of a Lyapunov based PD controller for position control. Tip position control of a TLFM is proposed [52] using a hybrid sliding mode consisting of frequency shaped optimal sliding mode control (FSOSMC) and terminal SMC (TSMC). Adaptive variable structure control is designed to estimate the upper bounds on the norm of uncertainties. The dead zone scheme is also introduced to improve the system robustness. Tip trajectory tracking control of a TLFM is presented in [53]. Separately an adaptive and SMC are designed for the control problem. The model of the manipulator is derived using the FEM method. Adaptive augmented state feedback control (AASFC) technique is designed in [54] for tip position and trajectory tracking control of a TLFM. The control in [54] consists of a steady state LQR technique in conjugation with an adaptive compensator, and STR (Strong Tracking Filter) for estimating the state. Tip position control of a TLFM handling different payloads using nonlinear adaptive model predictive control (NAMPC) is proposed in [4]. The ARMAX technique is used for system identification. The NAMPC performance is compared with the self tuning controller (STC) and direct adaptive controller (DAC). Reference [4] considered the tip trajectory tracking and vibration suppression as the performance measure. AMM is used for modelling the dynamics of the manipulator. Tip trajectory tracking and tip deflection suppression of a TLFM with variable payloads is proposed in [55] using adaptive and reinforcement learning (RL) control techniques. PD based adaptive controller and an actor-critic-based RL are used. Recursive Least Squares (RLS) based temporal difference (TD) learning is used for estimating the critic weights, and gradient based estimator for estimating actor weight. The proposed controller is compared with direct adaptive control and fuzzy adaptive controller. AMM is used for modelling the dynamics of the manipulator. Position control of TLFM with variable payloads is reported in [56]. It is based on the adaptive and RL control schemes. The LQR based decentralized controller is designed for the decoupled system. In this chapter, RL is used to tune the LQR gain.

H_∞ **control:** It is a non modeled based robust control technique. The controller is robust to model uncertainty, inaccuracy and is consistent in performance, order reduction and is insensitive to parameter variation. The common design method for this controller is to construct the system $p(s)$ as in (2.10) by augmenting the original system $y = G(s)u$ with the weighting function w such that the system $z = F_l(P, K)w$.

$$\begin{pmatrix} z \\ y \end{pmatrix} = p(s) \begin{pmatrix} w \\ u \end{pmatrix} \tag{2.10}$$

H_∞ control is obtained by minimizing the H_∞ norm of the system $F_l(P, K)$. The model reference adaptive H_∞ control is designed in [57] to handle unknown input nonlinearities such as dead-zone and backlash. The control scheme is designed for a mixed parameter system composed of distributed parameter systems of the hyperbolic type (flexible arm) and lumped parameter system (motor control system). Adaptive control is used to estimate and compensate the input nonlinearities. H_∞ control is designed to regulate the effect of the spillover term. Trajectory tracking control and vibration suppression of a TLFM is reported by [14] using a feedback controller based on H_∞ loop-shaping design and a feedforward compensator. LMI based gain scheduled controllers are designed for trajectory tracking control of a TLFM in [58]. The performance of the controller is shown on a AMM modeled TLFM. Tip position control of a TLFM is presented by [59] using gain scheduled strictly positive real controller. The AMM method is used for modelling the manipulator dynamics. The linear SPR controller is optimized using the gain scheduled controller. The results are demonstrated using numerical simulations.

Backstepping control: This is a non model based robust control technique. Handling bounded disturbances and uncertainties are the main advantages of the backstepping control technique. The most accepted required form of the system for the backstepping control design is given in (2.11).

$$\begin{cases} \dot{x}_1 = f_1(x_1, \ldots, x_{n-1}) \\ \dot{x}_n = f_n(x_1, \ldots, x_{n-1}, x_n) + u \end{cases} \tag{2.11}$$

Here, (2.11) is in strict feedback form. x_n is the nth state and u is the desired control input.
Tip trajectory tracking control and vibration suppression of a TLFM is presented in [60] using the backstepping control technique. The performance of the proposed controller is compared with the PD controller. The motion equation of the manipulator is developed using the projection equation and Ritz expansion technique.

Disturbance observer based control: It is a model based robust control technique. Independent joint control without considering the coupling effect of the other link is the main advantage of this control technique. The disturbance observer uses the difference between the actual input torque and the inverse of the output of the nominal model as the equivalent disturbance applied to it. To cancel the actual disturbance, equivalent disturbance is fed back through a low pass filter. The general structure of a disturbance observer for a flexible manipulator can be described as follows:

$$\dot{\hat{\tau}}_d = -L\hat{\tau}_d + L(M(\theta, \delta)\begin{pmatrix} \ddot{\theta} \\ \ddot{\delta} \end{pmatrix} + N(\theta, \dot{\theta}, \delta, \dot{\delta}) - u \tag{2.12}$$

where $\hat{\tau}_d$ is the estimated disturbance torque, L is the observer gain matrix and M, N are obtained from the FMs dynamics. A disturbance observer is used in [61] to compensate the disturbance for the position control of a TLFM. The lumped parameter method is used for modelling TLFMs.

State observer and backstepping control: Trajectory tracking control of a linear flexible TLFM is presented in [62]. A linear flexible system is obtained using nonlinear decoupling feedback control. An extended state observer is used to estimate the system's nonlinear part and backstepping control is used for trajectory tracking control of the manipulator.

Observer based LQR and SMC control: A hybrid control scheme is used in [63] to control the position of a TLFM. The hybrid controller is designed using continuous nonsingular terminal SMC and observer based LQR control techniques. Singular perturbation is used to divide the FM dynamics into two parts: slow and fast dynamics subsystems. For the slow subsystem a continuous nonsingular terminal SMC is designed and for the fast subsystem an observer based LQR controller is designed. The observer is designed to estimate the flexible modes and the LQR is designed to stabilize the modes.

LQG (Linear quadratic gaussian): A simple LQG controller with a linear plant model is described in (2.13):

$$\dot{x} = Ax + Bu + w_d$$

$$y = cx + w_n \tag{2.13}$$

where w_d, w_n are disturbance (process noise) and measurement noises, respectively, and are stochastic with known statistical properties. The LQG control problem is to find the optimal control input u which minimizes the following performance index.

$$J = E\{lim_{T\to\infty} \int_0^T [x^T Qx + u^T Ru]dt\} \tag{2.14}$$

where Q, R are the positive semi-definite and positive definite matrix, respectively, and optimal control is $u = -K_r x$ with an optimal gain matrix K_r.

The LQG control is useful for controlling the robot manipulator when the dynamics are considered with noisy inputs. Hybrid position and force control TLFM techniques are presented in [64] using the reduced order and full order LQG control techniques. The results are presented using numerical simulations and experiments. Trajectory tracking control of a TLFM is presented in [65] using model based LQG and feedforward control. Feedback gain is determined using the LQG regulation theory. FEM is used for modelling the dynamics of the manipulator. Reference [66] presented the trajectory tracking control of a TLFM using a combination of different robust control techniques. LQG, LQG/LQR, LTR and 7th order LQG/LTR, 7th order reduced LTR, 4th order SANDY optimization techniques are the different control methods used. The lumped parameter method is used for modelling the manipulator dynamics.

4. Intelligent control

Sometimes, the model of a real process is not known accurately or is difficult to formulate properly. So, due to the lack of an accurate model, the design of a control system is difficult. This problem can be overcome by incorporating linguistic information from human experts. Fuzzy logic control, artificial neural networks, and genetic algorithms are types of intelligent control techniques.

Fuzzy logic control: A conventional control technique may not work satisfactorily when the dynamics of the manipulator are not known accurately or if the dynamics vary in different operating conditions. However, designers can handle such complexities based on their experiences. Thus, a controller is required to design based on the linguistic information from human experiences. Ease in design and implementation without knowledge of the accurate model are the main advantages of a fuzzy logic controller as noted in [67]. It is a non modeled based control technique. The general structure of a FLC consists of four components (i) rule base, (ii) inference mechanism, (iii) fuzzification inference, (iv) defuzzification inference. In this structure, e and $\dot{e}$ variables of the flexible manipulator can be used as inputs to the fuzzy controller. A FLC is used in combination with other classical and robust controls for different TLFM control problems.

Reference [68] presents a TLFM fuzzy logic controller. The model of the manipulator is separated into two parts using singular perturbation techniques. Separate fuzzy logic controllers are designed for slow and fast dynamics for tracking control and deflection suppression of the manipulator, respectively. Tip position control and vibration suppression of a TLFM are presented in [69] using FLC and the control technique is termed as Fuzzy model reference learning controller. The FLC is designed in two forms, direct FLC and coupled FLC. A hybrid control scheme is proposed in [70] for trajectory tracking control and vibration suppression of a TLFM. A PID control is used for trajectory tracking and a SMC is used for vibration suppression of the manipulator. Fuzzy logic is used along with the PID for trajectory tracking control. Fuzzy learning and direct adaptive control (FLAC) is used by [71] for tip trajectory tracking control of a TLFM. Trajectory tracking control and tip deflection suppression are achieved in the presence of varying payloads. AMM is used to model the dynamics. Numerical simulations are done to generate the results. Fuzzy terminal sliding mode control is introduced by [15] for position control of a TLFM. The dynamics of the manipulator is divided into two parts using the two-time scale property. A fuzzy SMC is used for controlling a slow subsystem. A LQR and a reduced order observer are used for controlling the fast subsystem. AMM is used for modelling the dynamics of the flexible manipulator.

Artificial Neural Network (ANN): It is difficult to design a controller without knowing system dynamics. In such cases, an ANN is more suitable. It is a non modeled based intelligent control technique. The efficiency in building controllers for unknown dynamical systems and handling partially defined systems are the main advantages of this control technique. Feedforward and feedback networks are two types of ANN. An inverse dynamic model of a plant is used to train the ANN network. In a direct learning algorithm, ANNs copy the inverse plant dynamic model from the inputs and outputs of the plant. In indirect learning algorithms, the ANN mimics the inverse dynamic structure of the plant. ANNs are used in combination with other classical and robust controls for various TFLM control issues. Hybrid collected NN based MPC and non-collected PID controllers are designed in [72] for trajectory tracking control of a TLFM. In this AMM is used to model the dynamics and NN is used for identification of the dynamics. NNMPC is used for motion trajectory tracking and PID is designed for vibration control. Payloads are varied to study the effect on the performances of the controller. An end-effector trajectory tracking control of a TLFM is presented in [73] using a visual sensor based NN. A computed torque (similar to feedback linearization) controller is designed for virtual rigid robots and vision feedback signals. The link deflections of the manipulators are also suppressed. Anti-windup saturation compensation is achieved in [74] during trajectory tracking control of a TLFM. A singular perturbation is used to divide the dynamics into: slow and fast subsystems. NNs are designed to control the saturation nonlinearity in slow systems and LQRs are designed to attenuate the vibration in fast systems. A self tuning control (STC) for a TLFM using neural networks is presented in [16]. The proposed ANN controller in [16] learns the gains of the PI controller. Joint angle tracking control and vibration suppression are considered to be the control problems. The system identification of the plant is achieved using the ARMAX (Auto-regressive moving average exogenous) model. Reference [75] deals with the identification of a TLFM using an adaptive time delay neural network (ATDNN). Trajectory tracking is achieved using the proposed identification technique. The dynamic recurrent neural network is used for the purpose. The selection criteria for selecting the fixed structural parameters of the dynamics are also provided. Adaptation laws for updating the adjustable parameters of the network are also discussed. The proposed neuro dynamics are selected in series and parallel for training. A comparative study of model based and model free controls of a TLFM are presented in [76]. Reference [76] considered an inverse dynamics control as a model based control technique. Energy based control and neuro adaptive control are the two model free control techniques used. Inverse dynamics based control is achieved using PID control. Energy based control is achieved using a PD type control scheme and stability analysis is conducted using the Lyapunov stability theory. In this case, a neuro adaptive recursive neural network is used. Position control, force control and vibration control are the three control problems considered in [76]. Hybrid neuro fuzzy tracking control of a TLFM is presented in [77]. The primary loop of the proposed controller contains FLC and a neural network in the secondary loop for compensating the coupling effect between rigid and flexible movements along the inter-link. A radial basis function neural network is used for the purpose. The performance of the proposed hybrid neuro fuzzy control is compared with that of PD adaptive and fuzzy logic controllers. Reference [78] proposed tip-position trajectory tracking control of a TLFM using ANN and H_∞ controls. The dynamics of the manipulator are decomposed into a slow and a fast subsystem using an AMM and singular perturbation theory. H_∞ control is applied to the fast subsystem and decomposition based control on the slow subsystem. To compensate the quasi static deflection, a neural network compensation algorithm is used. Tip position and trajectory tracking problems are considered in [79] using an advanced control strategy. The control strategy is based on the PDE and ODE models of the manipulator. PID and feedback linearization based controls are considered to be an enhanced control strategy. Gaussian ANN is used in [79] to train the parameters of the controller. Trajectory tracking control and vibration suppression of a TLFM are reported in [80] using a neuro-SMC (NSMC). A singular perturbation technique is used to divide the dynamics of the manipulator. A NSMC is designed on slow subsystems for trajectory tracking and a normal SMC is designed for flexible subsystems to suppress the link vibrations.

Genetic algorithm (GA): This is a non-modeled based intelligent control technique. It uses a search procedure based on the mechanism of natural selection. It belongs to the class of heuristic methods and considered as a stochastic search algorithm. GA search for the best possible solution to optimization problems by mimicking the genetic dynamics of natural evolution. GA methods can be used either off-line or on-line. GA is used in combination with other classical and robust controls for different TLFM control problems. Tip position control of a TLFM is presented in [81] using an evolutionary computing based PD control technique. The evolutionary computing technique consists of a genetic algorithm and bacteria foraging optimization is used for optimizing the PD controller parameters. Terminal SMC control is proposed in [17] to represent the non-minimum phase characteristic using the output redefinition matrix of a TLFM. Manipulator dynamics are decomposed into an input-output subsystem and zero dynamics subsystem using the input-output linearisation technique. To control the input-output subsystem, a SMC is designed. The relation between the eigenvalue of the zero dynamics and parameters of the refined output is also obtained. To show the stability of the zero dynamics and entire manipulator dynamics, parameters of the controller are optimized using GA. Higher order nonsingular terminal SMC (HONTSMC) is designed in [82] to control the position of a TLFM in the presence of uncertainties. A third order NTSMC is designed for the stabilization of the input-output linearized dynamics in the absence of chattering. GA is used to optimize the controller gains. Optimized continuous NSTM is used in [83] for controlling the position of a TLFM. The controller is designed for an input-output subsystem to suppress the chattering and stabilize the dynamics. A GA is used for optimizing the parameters of the controller. Results are validated using simulations. An optimized continuous NSTM is used [84] for controlling the position of a TLFM in the presence of uncertainties and disturbances. A continuous NSTM is designed using SMC, and controller gains are optimized using GA techniques. The input-output subsystem is obtained to design the controller. The results are shown using simulation. Tip position and vibration suppression control for a TLFM using GAs are proposed in [85] based on a hybrid FLC strategy. An uncoupled FLC is used along with an individual controller at the shoulder and elbow link of the manipulator. A GA is used to extract the rule base of the FLC. A scaling factor of the FLC is tuned with GA to improve the performance. The FEM method is used to model the manipulator dynamics. A hybrid fuzzy non singular terminal SMC (NSTM) along with GAs is proposed for tip-position control of an uncertain TLFM [86]. The dynamics are decomposed into input-output and zero dynamics subsystems using an output redefinition method. Fuzzy NSTM is designed for input-output subsystems and Fuzzy is used to reduce the chattering in SMCs. The zero dynamics subsystem is controlled using a GA. Lyapunov stability theory is used for stability analysis of the system and AMM is used for manipulator dynamics modelling. Tip position control of a TLFM is proposed in [87] using GA optimized feedback and adaptive control. First dynamic state feedback control is used for suppressing vibrations and regulation of rigid modes. To compensate the payload changes and external disturbances, feedback control is used along with adaptive control. The controller gain is optimized using GA. An AMM is used for modelling the manipulator dynamics. Optimal trajectory control of a PDE modeled TLFM is presented in [88] using feedback control and GA. The optimal trajectory is generated using differential evaluation GA for minimizing the total energy consumption. Feedback boundary control is used to regulate the link in the optimized trajectory and to suppress the vibration in the manipulator links.

5. Hybrid control technique

Now a days, single control technique often insufficient to achieve the accuracy of the desired solution. Combinations of various control techniques are used to solve the problem. Hybrid control schemes, such as classical control combined with intelligent control, and robust control combined with intelligent control, are considered to enhance control performance. **Impulsive control:** Trajectory tracking control of a TLFM is presented in [89] using impulsive force moment based control. The proposed control strategy also works for suppression of elastic vibrations. The Lyapunov stability theory is used for stability analysis of the proposed scheme.

Lyapunov stability with other control scheme: Trajectory tracking control of a PDE modeled TLFM is reported in [90]. The manipulator dynamics are divided into slow sub and fast subsystems using the two-time scale method. The stability analysis of the fast subsystem for vibration suppression is accomplished using Lyapunov stability theory. Additionally, a PD type inverse dynamic based control is designed to ensure trajectory tracking of the slow subsystem. The proposed method in [90] excludes the observation and control spillover instability. The position control for a PDE modeled TLFM using passivity and Lyapunov based control is proposed in [91].

Trajectory tracking and vibration suppression of a TLFM using hybrid VSC and Lyapunov based control is reported in [92]. The equation of motion of a TLFM along with the PZT attached sensor and actuator is obtained using AMM and Hamilton's principle. The singular perturbation technique is used to divide the dynamics: slow and fast subsystem. A VSC is designed for trajectory tracking control of the slow subsystems. A Lyapunov based controller is designed for vibration suppression of fast subsystems.

[93] reported the position and vibration control of a TLFM using Lyapunov based control. The model of the manipulator dynamics is derived using the extended Hamiltonian's principle in coupled partial differential equations and ordinary differential equation forms with sufficient boundary conditions. The proposed modelling approach is compared with the FEM.

VSC and virtual force control: Trajectory tracking control of a TLFM is proposed in [94] using the hybrid control technique. A separate control law is designed for slow and fast subsystems. VSC is designed for trajectory tracking control in slow subsystems and VSC along with virtual control force is designed for suppressing the vibration and controlling the dynamics in closed loops for flexible subsystems. AMM is used for modelling the manipulator dynamics.

Generalized canonical transformation: Position control of a TLFM using the energy based generalized canonical transformation technique is reported in [95]. The dynamics of TLFMs are expressed as port controlled Hamiltonian systems. The performance of the proposed control is compared with PD control.

State feedback and dynamic extension technique: Tip position control of a TLFM with Kelvin-Voigt damping is proposed in [96]. Hamiltonian principle is used for modelling the dynamics using a set of coupled PDE and ODE. Output feedback control is designed using a dynamic extension method with a potential energy shaping technique. Local stability analysis is presented for the closed loop system.

Feedforward and state feedback using optimal strictly positive real (SPR) control: Tip position control of a single link and TLFM is presented in [97] using optimal strictly positive real (SPR) based feed-forward and feedback control. The SPR controller integrates rate (velocity) and position control approaches. In this an objective function is considered for minimizing the H_2 norm of closed loop dynamics.

PD, feedforward and state feedback control: Different control techniques are used by [98] for vibration suppression and trajectory tracking control of a TLFM. First collocated PD control is used for the control of rigid body motion, then hybrid control is designed for control of rigid body motion and vibration suppression using collocated PD and feed-forward control schemes. The feedforward control scheme is based on input shaping and a low pass filtering technique. Direct strain feedback control is designed for vibration control.

Model based inversion control and Lyapunov method: The robustness aspect of a model based controller used for trajectory tracking control of a TLFM is presented in [99]. The model is developed using the FEM method. A stage controller is designed for the robust control of a TLFM. In the first stage a second Lyapunov method, i.e., model based inversion control is developed for trajectory tracking control and in the second stage state feedback using end point sensing is developed for vibration suppression of the links. The performances of the controller are validated using numerical simulations.

Adaptive cerebellar model articulation controller: Reference [100] reported the position and vibration control of a TLFM using an adaptive cerebellar model articulation controller. AMM is used to obtain the dynamic manipulator model. The performance of the proposed control techniques is validated using numerical simulations.

2.2 Singular perturbation based controllers for TLFMs

In the late 1960's, the introduction the time scale and singular perturbation techniques were introduced in the field of control engineering. Since then, the singular perturbation has become an important tool for the design of control systems, analysis and modelling. Significant diversification is seen in recent publications which are growing in numbers in the field of control applications. The categories of development can be divided into three types. The first type includes the use of singular perturbation in control system oriented problems. The second type is in the refinement and extensions of theoretical concepts. The third type pertains to application-oriented and diverse specific problems, particularly in robotic manipulators. The dynamics of the manipulator is divided separately into two by the singular perturbation into slow and the fast subsystems. The slow subsystem corresponds to the rigid dynamics and the fast subsystem to the flexible transients. Singular perturbations are also used in the field of flight dynamics, manufacturing systems and chemical kinetics.

For control engineers, singular perturbation aids in addressing the high frequency phenomena that are often neglected by considering them as fast time-scale dynamics. This phenomenon is achieved by changing the dynamic order of a system of differential equations through a parameter perturbation that is more abrupt than a regular perturbation, a process referred to as a singular perturbation. The practical advantages of such a "parameterisation" of changes in model order are significant, because the order of every real dynamic system is higher than that of the model used to represent the system. This time scale separation also eliminates the difficulties in the stiffness and prepares for an efficient hardware and software implementation of the proposed controller.

There are few papers that deal with the implementation of singular perturbations on two-link flexible manipulators. Paper [101] deals with the two-link flexible manipulator control design using singular perturbations. In this paper, the fast subsystem is termed as inner-loop and is in O(1) order. These inner-loop dynamics are feedback linearised and the vibrations induced are taken into account. In the second stage, the slow subsystem is referred to as outer-loop controller meant for the suppression of vibrations due to rigid body motion. This controller is also used to enhance the robustness due to parameter variation in the system. The simulation results are performed and validation is done through experimentation. In [102], simulation and experimental validation are performed. The fast subsystem is designed with a PD based feedback control to deal with the frequency vibrations in O(1) order derived from asymptotic expansions. The slow subsystem is designed with a PD based feedback controller incorporated with an input preshaper. In this paper, it is shown that the applied control techniques significantly reduce the vibrations due to the geometric configurations of the manipulator link. It also cancels the nonlinearities developed due to the Coriolis and Centripetal effects. In the case of the construction of a huge space structure in the space station, control of vibrations and joint positions are not the only factors to be dealt with but also the force exerted by hands on the object. Paper [103] deals with this aspect where the problem of dynamic hybrid position/force control is taken into account. The elastic deflection of the flexible manipulator link is approximated by using the B-spline function and the Lagrange multiplier method for the dynamic equations of the joint angles in [103]. Though the manipulator has two links, the first link is rigid and the second is flexible in nature. A composite controller is designed where the PID feedback controller is designed for the fast subsystem and the PID controller is designed for the slow subsystem. Hence, the proposed controller and the model in [103] is useful for the constraint flexible manipulator. A reduced order model of a two-link flexible manipulator with flexible joints is modelled through AMM along with singular perturbations. A composite controller is designed to resolve the complexities of control associated with under-actuated flexible links and joints. The slow subsystem controller is designed using computed torque control and the fast subsystem controller is designed using LQR based statefeedback control. The proposed controller maintains a good tracking performance as well as to stabilise the links and the joints [104]. The proposed control schemes used for the subsystems are VSC (variable structure controls) and Virtual force controls for slow and the fast subsystems, respectively [105]. This paper suggests that the proposed control scheme is more robust in tracking than the conventional scheme in the presence of the payload. In [106], a new control scheme is proposed for the two-link flexible manipulator that is an energy-based two-time scale control design. The modelling is accomplished using Euler-lagrange with energy reshaping and damping injection, segregating into two time scales subsystems. PD controllers and feedback controllers are designed for the slow and the fast subsystems, respectively. The effectiveness of the controller is experimentally validated. A robust control method for a two-link flexible manipulator with neural networks based quasi-static distortion compensation is proposed and experimentally implemented in [107]. In [108], a composite controller is designed on the basis of singular perturbations for a two-link flexible manipulator with external disturbances and parameter uncertainties. A new adaptive sliding mode control and an optimal control are proposed for the slow and the fast subsystems respectively. The use of this proposed controller significantly reduces the chattering as compared to the conventional sliding mode with tip deflection suppression. A fuzzy terminal sliding mode control method is employed for the control of two-link flexible manipulator for the slow subsystem, and LQR is utilized for the fast system [109]. The simulation is done to validate the robustness of the proposed controller. A PD type inverse dynamic based controller and a Lyapunov based controller are designed for the slow and fast subsystems, respectively [110]. This paper depicts that the proposed method does not necessitate any information about the vibration of the links along the links for the proposed fast subsystem and the discretisation of the PDE of the arm vibration to the set of ODE's. Hence, these methods exclude the effect of both the observation and control spillover instability. A normal SMC and H_∞ controllers are designed for the slow and the fast subsystems, respectively in order to overcome the non structural uncertainties caused by the unmodelled system dynamics [111]. The experimental results are conducted and compared with a standard PID controller to validate the robustness of the proposed controller. Paper [92] addresses vibration suppression and maneuver control of a two-link flexible manipulator arm utilizing embedded smart materials. The equations of motion of the dynamics with PZT patches and sensors are obtained. A hybrid controller is proposed for stability using the Lyapunov based approach. A Variable Structure Controller (VSC) is proposed for the slow subsystem while a Lyapunov based controller is proposed for the fast subsystem to enhance maneuver control efficiency and suppress vibrations in the structures. An anti-windup saturation compensation scheme for a two-link flexible manipulator is studied in [74]. The nonlinear system is then decomposed into slow and fast subsystems using singular perturbation. A neural network is designed to compensate for the nonlinear saturation in rigid subsystem, while a robust controller is used for the fast subsystem. In [112], a continuous nonsingular terminal sliding mode controller is proposed for the slow subsystem, and an observer based LQR controller is proposed for the slow subsystem. Simulation results demonstrate the effectiveness of the proposed controllers.

2.3 Tracking controllers of TLFMs for a desired path as a chaotic signal

Many control techniques have been reported for the trajectory tracking of a two-link flexible manipulator. However, the utilization of a chaotic signal as a desired trajectory for a TLFM is a novel approach. Some papers in the field of mobile robots and synchronisation of a flexible arm with an electromechanical device have utilized chaotic signals. For instance, Paper [113] focuses on the synchronisation between a nonlinear, self sustained chaotic oscillator (the electrical part) and a mechanical part with a flexible beam. Additionally [114], a chaotic path planning generator for mobile robots is presented. This paper focuses the controller based generator that defines the position goal in each step by imparting chaotic motion behaviour. Numerical simulations conducted on robot motion control validate that the proposed technique can yield satisfactory results with respect to unpredictability and rapid scanning of the robot's workplace. In [115], a chaotic motion path planner utilizing a Logistic Map is introduced for an autonomous mobile robot. The aim is to cover an unknown, random, and unpredictable terrain. The enhanced path planner ensures that the chosen closed contour exhibits sensitive dependence on initial conditions, to meeting the needs of an autonomous patrolling robot. Paper [116] represents a path planning algorithm for every autonomous mobile robot based on the chaotic dynamics of the Hénon system. The system is chosen to ensure the unpredictable boundary patrol on any chosen contour.

2.4 Synchronisation control algorithms of TLFMs

Synchronisation/cooperation between manipulators is a specified emerging application in the field of robotics. However, most of the synchronisation works are reported using rigid manipulators. Flexible manipulators have their advantages over rigid manipulators and hence, their applications are more desirable.

Many papers are available on the synchronisation of rigid robot manipulators [117], [118], [119–121]. The synchronisation of robot manipulators can be broadly divided into three categories. These are motion synchronisation [122], [123], [121], force synchronisation [119], [121], [124] and task space/ trajectory synchronisation [125, 126]. The synchronisation of robot manipulators are categorically explained in [123], [120]. It is to be noted that most of the available synchronisation concepts are focused on the rigid manipulators which are generally not precise, economical and effective in industrial applications.The synchronisation between two-link flexible manipulators (TLFM) is a new approach in industrial applications. Hence, the synchronisation between two two-link flexible manipulators is rarely available in the literature.

2.5 Chapter summary

In this chapter, broad and general literature surveys on different sections of the problems considered in the book are presented. The literature survey is broadly divided into four sections. These are (i) on control techniques, (ii) singular perturbation based controller, (iii) tracking control for chaotic desired signal and (iv) synchronisation control algorithm for the TLFMs. Thus, based on these surveys, objectives of the book are formulated.

References

[1] F. Wang and Y. Gao. Advanced Studies of Flexible Robotic Manipulators, Modeling, Design, Control and Applications. *World Scientific*, vol. 5, no. New Jersey, pp. ISBN: 978–981–279–672–1, 2003.

[2] F. Matsuno and K. Yamamoto. Dynamic Hybrid Force/Position Control of a Two Degree-of-freedom Flexible Manipulator. *Journal of Robotic Systems*, vol. 11, pp. 355–366, March 2007.

[3] W. Chen, Y. Yu, L. Zhao, and Q. Sun. Position Control of a 2DOF Underactuated Planar Flexible Manipulator. In *International Conference on Mechatronics and Automation (ICMA)*, (Beijing), pp. 464–469, August 2011.

[4] S. K. Pradhan and B. Subudhi. Nonlinear Adaptive Model Predictive Controller for a Flexible Manipulator: An Experimental Study. *IEEE Transactions on Control Systems Technology*, vol. 22, pp. 1754–1768, September 2014.

[5] A. Parida and S. Ranasingh. *Modelling And Robust PD Compensation of Two-Link Flexible Manipulator*. Thesis, NIT Rourkela, 2011.

[6] F. Khorrami and J. Sandeep. Experiments on Rigid Body-Based Controllers with Input Preshaping for a Two-Link Flexible Manipulator. *IEEE Transactions on Robotics and Automation*, vol. 10, pp. 55–65, February 1994.

[7] F. L. Hu and A. G. Ulsoy. Dynamic Modeling of Constrained Flexible Robot Arms for Controller Design. *Journal of Dynamic Systems Measurement and Control*, vol. 116, no. 1, pp. 336–343, 1994.

[8] A. Bazaei and M. Moallem. Improving Force Control Bandwidth of Flexible-Link Arms Through Output Redefinition. *IEEE/ASME Transactions on Mechatronics*, vol. 16, pp. 380–386, April 2011.

[9] H. Alazki, P. Ordaz, and A. Poznyak. Robust Bounded Control for the Flexible Arm Robot. *Proc. of IEEE Conference on Decision and Control*, pp. 3061–3066, 2013.

[10] S. Cetinkunt and W. J. Book. Optimum Control of Flexible Robot Arms on Fixed Paths. In 3rd *Army Conference on Applied Mathematics and Computing*, pp. 1–12, February 1986.

[11] M. Mosayebi and M. Ghayour. Observer Based Tip Tracking Control of Two-link Flexible Manipulator. In *International Conference on Control Automation and Systems (ICCAS)*, (Gyeonggi-do), pp. 9–13, October 2010.

[12] A. Sanz and S. Etxebarria. Experimental Control of a Two-DOF Flexible Robot Manipulator by Optimal and Sliding Methods. *Journal of Intelligent and Robotic Systems*, vol. 46, pp. 95–110, August 2006.

[13] J. J. E. Slotine and L. Weiping. Adaptive Manipulator Control: A Case Study. *IEEE Transactions on Automatic Control*, vol. 33, pp. 995–1003, November 1998.

[14] M. Sayahkarajy, Z. Mohamed, and A. A. M. Faudzi. Review of modelling and control of flexible-link manipulators. *Journal of Systems and Control Engineering*, pp. 1–13, 2016.

[15] Y. Wang, Y. Feng, and X. Yu. Fuzzy Terminal Sliding Mode Control of Two-link Flexible Manipulators. In 34th *Annual Conference of IEEE Industrial Electronics (IECON)*, (Orlando, FL), pp. 1620 – 1625, November 2008.

[16] M. Sasaki, A. Asai, and T. Shimizu. Self-Tuning Control of a Two-Link Flexible Manipulator using Neural Networks. In *ICCAS-SICE, 2009*, (Fukuoka), pp. 2468 – 2473, August 2009.

[17] N. Zhang, Y. Feng, and X. Yu. Optimization of Terminal Sliding Control for Two-link Flexible Manipulators. In 30th *IEEE annual Conference of the Industrial Electronics Society*, vol. 2, (Busan, Korea), pp. 1318–1322, November 2004.

[18] S. Karande, P. S. V. Nataraj, and P. S. Ghandhi. Control of Parallel Flexible Five Bar Manipulator using QFT. In *IEEE International Conference on Industrial Technology*, (Gippsland, VIC), pp. 1–6, February 2009.

[19] R. M. Mahamood and J. J. Pedro. Hybrid PD/PID Controller Design for Two-Link Flexible Manipulators. In 8th *Asian Control Conference (ASCC)*, (Kaohsiung), pp. 1358 – 1363, May 2011.

[20] A. Walsh and J. Richard Forbes. Modeling and Control of Flexible Telescoping Manipulators. *IEEE Transaction on Robotics*, vol. 31, no. 4, pp. 936–947, 2015.

[21] C. Qingsong and Y. Ailan. Optimal Actuator Placement for Vibration Control of Two-link Piezoelectric Flexible Manipulator. *International Conference on Mechanic Automation and Control Engineering*, no. 2, pp. 2448–2451, 2010.

[22] Z. Mohamed, M. Khairudin, A. R. Husain, and B. Subudhi. Linear Matrix Inequality-based Robust Proportional Derivative Control of a Two-link Flexible Manipulator. *Journal of Vibration and Control*, no. June, pp. 1–12, 2013.

[23] M. Sawada and K. Itamiya. A Position Control of 2 DOF Flexible Link Robot Arms Based on Computed Torque Method. In *Proceedings of the 2012 International Conference on Advanced Mechatronic Systems*, pp. 547–552, 2012.

[24] A. Jnifene and A. Fahim. Endpoint Control of a Two-link Flexible Manipulator. *Journal of Vibration and Control*, vol. 4, no. 6, pp. 747–766, 1998.

[25] A. S. Morris and A. Madani. Computed Torque Control Applied to a Simulated Two-flexible-link Robot. *Transactions of the Institute of Measurement and Control*, vol. 19, no. 1, pp. 50–60, 1997.

[26] W. J. Book, O. Maizza-Neto, and D. E. Whitney. Feedback Control of Two Beam, Two Joint Systems with Distributed Flexibility. *Journal of Dynamical System Measurement and Control*, vol. 97, pp. 424–431, December 1975.

[27] O. M. Neto. Nodal Analysis and Control of Flexible Manipulator Arms," *PhD Thesis*, 1974.

[28] M. Benosman, G. Le Vey, L. Lanari, and A. De Luca. Rest-to-Rest Motion for Planar Multi-Link Flexible Manipulator Through Backward Recursion. *Journal of Dynamic Systems, Measurement, and Control*, vol. 126, no. 1, p. 115, 2004.

[29] T. Fukuda and A. Arakawa. Modeling and Control Characteristics for a Two-Degree-of Freedom Coupling System of Flexible Robot Arm. *JSME international Journal*, vol. 30, no. 267, pp. 1458–1464, 1987.

[30] H. Asada, Z. D. Ma, and H. Tokumaru. Inverse dynamics of flexible robot arms: Modeling and computation for trajectory control. *Journal of dynamic systems*, vol. 112, pp. 177–185, 1990.

[31] Y. Aoustin and A. Formalsky. On the Feedforward Torques and Reference Trajectory. *Multibody System Dynamics*, vol. 3, pp. 241–265, August 1999.

[32] X. Cao and Y. Li. Distributed Parameter Singular Perturbation Model and Cooperative Control of Flexible Manipulators. In *Proc. of International Conference on Machine Learning and Cybernetics*, vol. 2, (Guangzhou, China), pp. 1009–1014, August 2005.

[33] A. Vandini, A. Salerno, C. J. Payne, and G. Yang. Vision-Based Motion Control of a Flexible Robot for Surgical Applications. *IEEE International Conference on Robotics and Automation (ICRA)*, pp. 6205–6211, 2014.

[34] Z. Chu and J. Cui. Experiment on Vibration Control of A Two-link Flexible Manipulator using An Input Shaper and Adaptive Positive Position Feedback.' *Advances in Mechanical Engineering*, vol. 7, no. 10, pp. 1–13, 2015.

[35] F. Khorrami and J. Sandeep. Nonlinear Control with End-Point Acceleration Feedback for a Two-Link Flexible Manipulator: Experimental Results. *Journal of Robotic Systems*, vol. 10, pp. 505–530, June 1993.

[36] L. Zhang and J. Liu. Observer Based Partial Differential Equation Boundary Control for A Flexible Two-Link Manipulator in Task Space . *IET Control Theory and Application*, vol. 13, pp. 2120–2133, 2012.

[37] L. Zhang and J. Liu. Nonlinear PDE Observer Design for a Flexible Two-Link Manipulator. In *American Control Conference (ACC)*, (Montreal, QC), pp. 5336–5341, June 2012.

[38] W. Yim. Cartesian Trajectory Control of a Flexible Manipulator Using Sliding Mode. *Mechatronics*, vol. 4, pp. 635–652, September 1994.

[39] W. Dongmei. The Design of Terminal Sliding Controller of Two-Link Flexible Manipulators. In *IEEE International Conference on Control and Automation, ICCA*, (Guangzhou), pp. 733–737, June 2007.

[40] K. Kherraz, M. Hamerlain, and N. Achour. Robust Sliding Mode Controller for a Class of Under-actuated Systems. In 15th *International Conference on Sciences and Techniques of Automatic Control & Computer Engineering*, (Hammamet), pp. 942–946, 2014.

[41] R. Fareh and M. Saad. Tracking Control of Two-flexible-link Manipulators Using Distributed Control Strategy. In *2013 International Conference on Control Decision and Information Technologies (CoDIT)*, (Hammamet), pp. 653–658, 2013.

[42] H. C. Shin and S. B. Choi. Position Control of a Two-link Flexible Manipulator Featuring Piezoelectric Actuators and Sensors. *Mechatronics*, vol. 11, pp. 707–729, September 2001.

[43] S. Yurkovich, K. J. Hillsley, and A. P. Txes. Identification and Control For a Manipulator With Two Flexible Links. In *Proc. of the* 29th *IEEE Conference on Decision and Control*, vol. 4, (Honolulu, HI), pp. 1995–2000, December 1990.

[44] S. Yurkovich, A. P. Txes, I. Lee, and L. Hillsley. Control and System Identification of a Two-Link Flexible Manipulator. In *Proc. of IEEE International Conference on Robotics and Automation*, vol. 3, (Cincinnati, OH), pp. 1626 – 1631, May 1990.

[45] L. Zhang and J. Liu. Adaptive Boundary Control for Flexible Two-link Manipulator Based on Partial Differential Equation Dynamic Model. *IET Control Theory Appllication*, vol. 7, pp. 43–51, January 2013.

[46] M. Sawada and K. Itamiya. Adaptive Position Control with Using a Smooth Projection Adaptation Law for 2 DOF Flexible Link Robot Arm. *SICE Annual Conference*, pp. 1210–1215, 2012.

[47] M. Bai, D. H. Zhou, and H. Schwarz. Adaptive Augmented State Feedback Control for An Experimental Planar Two-link Flexible Manipulator. *IEEE Transactions on Robotics and Automation*, vol. 14, no. 6, pp. 940–950, 1998.

[48] R. I. Milford and S. F. Asokanthan. Identification and Gain Scheduled Vibration Control of An Experimental Two-link Flexible Manipulator. In *American Control Conference,*, pp. 3326–3328, 1995.

[49] J. Liu and L. Zhang. Adaptive Boundary Control for Flexible Two-link Manipulator Based on Partial Differential Equation Dynamic Model. *IET Control Theory & Applications*, vol. 7, no. 1, pp. 43–51, 2013.

[50] S. Ozcelik and E. Miranda. *Output Feedback Adaptive Control for a Two-Link Flexible Robot Subject to Parameter Changes in Adaptive Control.* I-Tech Publishers, 2008.

[51] S. Hosseini, A. Fallah, and M. Nazari. Adaptive Energy-Based Controllers for a Two Link Flexible Manipulator under Gravity. In *International Conference on Control Automation and Systems (ICCAS)*, (Seoul), pp. 659–663, October 2008.

[52] W. Cao and J. Xu. Dynamic Modeling and Adaptive VSC of Two-link Flexible Manipulators using a Hybrid Sliding Surface. In *Proceedings of the* 39th *IEEE Conference on Decision and Control*, vol. 5, (Sydney, Australia), pp. 5143–5148, December 2000.

[53] Z. Jiang. Impedance Control of Flexible Robot Arms with Parametric Uncertainties. *Journal of Intelligent and Robotic Systems*, vol. 42, no. 2, pp. 113–133, 2005.

[54] M. Bai, D. H. Zhou, and H. Schwarz. Adaptive Augmented State Feedback Control for an Experimental Planar Two-Link Flexible Manipulator. *IEEE Transactions on Robotics and Automation*, vol. 14, pp. 940–950, December 1998.

[55] S. K. Pradhan and B. Subudhi. Real-Time Adaptive Control of a Flexible Manipulator Using Reinforcement Learning. *IEEE Transactions on Automation Science and Engineering*, vol. 9, pp. 2012, 273–249, April 2012.

[56] B. Subudhi and S. K. Pradhan. Direct Adaptive Control of a Flexible Robot using Reinforcement Learning. In *IEEE International Conference on Industrial Electronics, Control and Robotics*, (Orissa, India), pp. 129–136, 2010.

[57] Y. Miyasato. Finite Dimensional Adaptive H_∞ Control for Flexible Arms Preceded by Input Nonlinearites. In *IEEE International Symposium on Intelligent Control Part of* 2010 *IEEE Multi-Conference on Systems and Control*, (Yokohama, Japan), pp. 2296–2301, 2010.

[58] P. Apkarian and R. J. Adams. Advanced Gain-scheduling Techniques for Uncertain Systems. *IEEE Transactions on Control Systems Technology*, vol. 6, no. 1, pp. 21–32, 1998.

[59] J. Forbes and C. Damaren. Design of Gain-Scheduled Strictly Positive Real Controllers Using Numerical Optimization for Flexible Robotic Systems. *Journal of Dynamic Systems, Measurement, and Control*, vol. 132, no. 3, p. 34503, 2010.

[60] P. Stauter, H. Gattringer, W. Höbart, and H. Bremer. Passivity Based Backstepping Control of an Elastic Robot. In *Robot Design, Dynamics and Control*, pp. 315–322, Springer Vienna, 2010.

[61] J. Cheong, W. Chung, Y. Youm, and S. Oh. Control of Two-Link Flexible Manipulator Using Disturbance Observer with Reaction Torque Feedback. In *Proc. of* 8th *International Conference on Advanced Robotics (ICAR)*, (Monterey, CA), pp. 227–232, 1997.

[62] H. Yang, Y. Yu, Y. Yuan, and X. Fan. Back-stepping Control of Two-link Flexible Manipulator Based on an Extended State Observer. *Advances in Space Research*, vol. 56, pp. 2312–2322, 2015.

[63] Y. Wang, F. Han, Y. Feng, and X. Hongwei. Hybrid Continuous Nonsingular Terminal Sliding Mode Control of Uncertain Flexible Manipulators. In 40th *Annual Conference of the IEEE IIndustrial Electronics Society (IECON)*, no. 51307035, (Dallas, TX), pp. 190–196, 2014.

[64] D. Bossertl, U. L. Juris, and J. Vegners. Experimental Comparison of Robust Reduced-Order Hybrid Position and Force Optimization TEchniques for a Two-Link Flexible Manipulator. In *IEEE international Conference on Control Applications*, (Dearborn), pp. 982–987, 1996.

[65] D. A. Schoenwald, J. A. Feddema, G. R. Eider, and D. A. Segalman. Minimum-time Trajectory Control of a Two-link Flexible Robotic Manipulator. In *Proc. of IEEE International Conference on Robotics and Automation*, vol. 3, (Sacramento, CA), pp. 2114–2120, April 1991.

[66] D. Bossert, U. Ly, and J. Vagners. Evaluation of Reduced-Order Controllers on a Two-Link Flexible Manipulator. In *Proc. of the American Control Conference*, vol. 5, (Seattle, WA), pp. 3339–3343, June 1995.

[67] L. Meirovitch. *Methods of Analytical Dynamics*. McGraw-Hill, 1970.

[68] L. Zhang, F. Sun, and Z. Sun. Cloud Model-based Controller Design for Flexible-link Manipulators. *IEEE Conference on Robotics, Automation and Mechatronics*, no. 1, 2006.

[69] V. G. Moudgal, K. M. Passino, and S. Yurkovich. Rule-Based Control for a Flexible-Link Robot. *IEEE Transaction on Control System Technology*, vol. 2, no. 4, pp. 392–405, 1994.

[70] M. A. Ali, F. B. Ismail, K. S. M. Sahari, and K. Weria. Trajectory Tracking Controller for Flexible Robot. In *IEEE International Symposium on Robotics and Manufacturing Automation*, (Kuala Lumpur), pp. 39–45, 2014.

[71] S. K. Pradhan and B. Subudhi. Fuzzy Learning based Adaptive Control for a Two-Link Flexible Manipulator. *Proc. of IEEE International Conference on Control Applications (CCA)*, pp. 282–287, 2013.

[72] J. O. Pedro and T. Tshabalala. Hybrid NNMPC / PID Control of a Two-Link Flexible Manipulator with Actuator Dynamics. In 10th *Asian Control Conference (ASCC)*, (Firenze), pp. 1–6, 2015.

[73] Z. Jiang. Workspace Trajectory Control of Flexible Robot Manipulators Using Neural Network and Visual Sensor feedback. In *Proceeding of the IEEE 28th Canadian Conference on Electrical and Computer Engineering*, vol. 1502, (Halifax, NS), pp. 1502–1507, 2015.

[74] D. Yue-jiao, C. Xi, Z. Ming, and R. Jun. Anti-windup for Two-link Flexible Arms with Actuator Saturation using Neural Network. In *2010 International Conference on E-Product E-Service and E-Entertainment (ICEEE),*, no. 06, pp. 6–9, 2010.

[75] A. Yazdizadeh, K. Khorasani, and R. A. Patel. Identification of a Two-Link Flexible Manipulator Using Adaptive Time Delay Neural Networks. *IEEE Transactions on Systems Man and Cybernetics*, vol. 30, pp. 165–172, February 2000.

[76] G. G. Rigatos. Model-based and Model-free Control of Flexible-link Robots: A Comparison Between Representative Methods. *Applied Mathematical Modelling*, vol. 33, pp. 3906–3925, October 2009.

[77] B. Subudhi and A. S. Morris. Soft Computing Methods Applied to the Control of a Flexible Robot Manipulator. *Applied Soft Computing*, vol. 9, pp. 149–158, January 2009.

[78] Y. Li, G. Liu, T. Hong, and K. Liu. Robust Control of a Two-Link Flexible Manipulator with Quasi-Static Deflection Compensation Using Neural Networks. *Journal of Intelligent and Robotic Systems*, vol. 44, pp. 263–276, November 2005.

[79] X. Zhang, W. Xu, and S. S. Nair. Comparison of Some Modeling and Control Issues for a Flexible Two Link Manipulator. *ISA Transaction*, vol. 43, pp. 509–525, October 2004.

[80] Y. Zhang, T. Yang, and Z. Sun. Neuro-Sliding-Mode Control of Flexible-Link Manipulators Based on Singularly Perturbed Model. *Tsinghua Science and Technology*, vol. 14, no. 4, pp. 444–451, 2009.

[81] B. Subudhi, S. Ranasingh, and A. K. Swain. Evolutionary Computation Approaches to Tip Position Controller Design for A Two-link Flexible Manipulator. *Archives of Control Sciences*, vol. 21, no. 3, pp. 269–285, 2011.

[82] Y. Wang, C. Wang, P. Lu, and Y. Wang. High-order Nonsingular Terminal Sliding Mode Optimal Control of Two-Link Flexible Manipulators. In 37th *IEEE Annual Conference on Industrial Electronics Society (IECON)*, (Melbourne, VIC), pp. 3953–3958, 2011.

[83] Y. Wang and L. Sun. On the Optimized Continuous Nonsingular Terminal Sliding Mode Control of Flexible Manipulators. 4th *International Conference on Instrumentation and Measurement Computer Communication and Control*, pp. 324–329, 2014.

[84] Y. Wang, C. Yuqing, and X. Hongwei. Optimized Continuous Nonsingular Terminal Sliding Mode Control of Uncertain Flexible Manipulators. In *Proc. of* 34th *Chinese Control Conference*, (Hangzhou), pp. 3392–3397, 2015.

[85] T. Zebin and M. S. Alam. Dynamic Modeling and Fuzzy Logic Control of a Two-link Flexible Manipulator using Genetic Optimization Techniques. In *Proc. of* 13th *IEEE International Conference on Computer and Information Technology (ICCIT)*, (Dhaka), pp. 418–423, December 2010.

[86] Y. Wang, H. Xia, and C. Wang. Hybrid Controllers for Two-Link Flexible Manipulators. *Applied Informatics and Communication in Computer and Information Science*, vol. 226, pp. 409–418, 2011.

[87] M. Dogan and Y. Istefanopulos. Optimal Nonlinear Controller Design for Flexible Robot Manipulators with Adaptive Internal Model. *IET Control Theory and Applications*, vol. 1, pp. 770–778, May 2007.

[88] L. Zhang and J. Liu. Optimal Trajectory Control of Flexible Two-link Manipulator Based on PDE Model. In 51st *IEEE Conference on Decision and Control (CDC)*, (Maui, HI), pp. 4406–4411, 2012.

[89] L. Xiaoguang and L. Junyi. A Controller with Impulse Force Moment Design for Two-Links Flexible Manipulator. In *Proc. of* 4th *World Congress on Intelligent Control and Automation*, vol. 4, pp. 3058 – 3062, 2002.

[90] A. Ashayeri, M. Farid, and M. Eghtead. Trajectory Tracking for Two-link Flexible Arm via Two-time Scale and Boundary Control Methods. In *Proc. of IMECE, ASME International Mechanical Engineering Congress and Exposition*, (Massachusetts, USA), pp. 189–197, 2008.

[91] X. Zhang, W. Xu, S. S. Nair, and V. Chellaboina. PDE Modeling and Control of a Flexible Two-Link Manipulator. *IEEE Transaction on Control systems Technology*, vol. 13, pp. 301–312, March 2005.

[92] E. Mirzaee, M. Eghtesad, and S. A. Fazelzadeh. Maneuver Control and Active Vibration Suppression of A Two-link Flexible Arm using A Hybrid Variable Structure/Lyapunov Control Design. *Acta Astronautica*, vol. 67, no. 9-10, pp. 1218–1232, 2010.

[93] M. Dogan and O. Morgul. On the Control of Two-Link Flexible Arm with Nonuniform Cross Section. *Journal of Vibration and Control*, vol. 16, no. 5, pp. 619–646, 2010.

[94] S. Lee and C. Lee. Hybrid Control Scheme for Robust Tracking of Two-Link Flexible Manipulator. *Journal of Intelligent and Robotic Systems*, vol. 34, pp. 431–452, July 2002.

[95] X. Bo, Y. Fujimoto, and Y. Hayakawa. Control of Two-link Flexible Manipulators via Generalized Canonical Transformation. In *IEEE Conference on Robotics Automation and Mechatronics*, vol. 1, pp. 107–112, December 2004.

[96] T. Shimizu, M. Sasaki, and T. Okada. Tip Position Control of a Two Links Flexible Manipulator Based on the Dynamic Extension Technique. In *SICE Annual Conference*, (Takamatsu), pp. 868–873, September 2007.

[97] J. R. Forbesa and C. J. Damaren. Design of Optimal Strictly Positive Real Controllers Using Numerical Optimization for the Control of Flexible Robotic Systems. *Journal of the Franklin Institute*, vol. 348, pp. 2191–2215, October 2011.

[98] Z. Mohamed, J. M. Martins, M. O. Tokhi, J. SadaCosta, and M. A. Botto. Vibration Control of a very Flexible Manipulator System. *Control Engineering Practice*, vol. 13, pp. 267–277, March 2005.

[99] R. J. Theodore and A. Ghosal. Robust Control of Multilink Flexible Manipulators. *Mechanism and machine theory*, vol. 38, no. 4, pp. 367–377, 2003.

[100] A. R. Maouche and H. Meddahi. A Fast Adaptive Artificial Neural Network Controller for Flexible Link Manipulators. *International Journal of Advanced Computer Science and Applications*, vol. 7, no. 1, pp. 298–308, 2016.

[101] F. Khorrami and S. Jain. Non-Linear Control with End-Point Acceleration Feedback for a Two-Link Flexible Manipulator: Experimental Results. *Journal of Robotic Systems*, vol. 10, no. 4, pp. 505–530, 1993.

[102] F. Khorrami, S. Jain, and A. Tzes. Experiments on Rigid Body-Based Controllers with Input Preshaping for a Two-Link Flexible Manipulator. *IEEE Transaction on Robotics and Automation*, vol. 10, no. 11, pp. 55–65, 1994.

[103] F. Matsuno and K. Yamamoto. Dynamic Hybrid Force/Position Control of Two Degrre of Freedom Flexible Manipulator. *Journal of Robotic Systems*, vol. 11, no. 5, pp. 355–366, 1994.

[104] B. Subudhi and A. S. Morris. Dynamic Modelling, Simulation and Control of A Manipulator with Flexible Links and Joints. *Robotics and Autonomous Systems*, vol. 41, no. 4, pp. 257–270, 2002.

[105] S. H. Lee and C. W. Lee. Hybrid Control Scheme for Robust Tracking of Two-link Flexible Manipulator. *Journal of Intelligent & Robotic Systems*, vol. 34, no. 4, pp. 431–452, 2002.

[106] X. U. Bo and Y. Bakakawa. Control Two-link Flexible Manipulators using Controlled Lagrangian Method. In *SICE Annual Conference in Sapporo*, pp. 289–294, 2004.

[107] Y. Li, G. Liu, T. Hong, and K. Liu. Robust Control of A Two-link Flexible Manipulator with Quasi-static Deflection Compensation using Neural Networks. *Journal of Intelligent & Robotic Systems*, vol. 44, no. 3, pp. 263–276, 2005.

[108] Y. Zhang, Y. Mi, M. Zhu, and F. Lu. Adaptive Sliding Mode Control For Two-Link Flexible Manipulator with H infinity tracking. No. August, pp. 702–707, 2005.

[109] Y. Wang, Y. Feng, and X. Yu. Fuzzy Terminal Sliding Mode Control of Two-link Flexible Manipulators, vol. 2, pp. 1620–1625, 2008.

[110] A. Ashayeri and M. Farid. Trajectory Tracking for Two-Link Flexible Arm via Two-Time Scale and Boundary Control Methods. In *Proceedings of IMECE2008 2008 ASME International Mechanical Engineering Congress and Exposition*, pp. 1–9, 2008.

[111] Y. C. Li, B. J. Tang, Z. X. Shi, and Y. F. Lu. Experimental Study for Trajectory Tracking of A Two-link Flexible Manipulator. *International Journal of Systems Science*, vol. 31, no. 1, pp. 3–9, 2010.

[112] Y. Wang, F. Han, Y. Feng, and X. Hongwei. Hybrid Continuous Nonsingular Terminal Sliding Mode Control of Uncertain Flexible Manipulators. In *40th IEEE Annual Conference of the Industrial Electronics Society (IECON)*, no. 51307035, pp. 190–196, 2014.

[113] C. A. K. Kwuimy and P. Woafo. Dynamics, chaos an dsynchronisation of Self-sustained Electromechanical systems with Clamped-free Flexible Arm. *Nonlinear Dynamics*, vol. 53, pp. 201–213, 2008.

[114] C. K. Volos, L. M. Kyprianidis, and S. I. N. A Chaotic Path Planning Generator for Autonomous Mobile Robots. *Robotics and Autonomous System*, vol. 60, pp. 651–656, 2012.

[115] C. Li, W. Fengjing, Z. Lei, and S. Yong. An Improved Chaotic Motion Path Planner for Autonomous Mobile Robots based on a Logistic Map. *International Journal of Advanced Robotic Systems*, vol. 60, pp. 1–7, 2012.

[116] D. Curiac and V. Constantine. A 2D chaotic path Planning for Mobile Robots Accomplishing Boundary Surveillance Missions in Adversarial Conditions. *Commun Nonlinear Sci Numer Simulat*, vol. 19, pp. 3617–3627, 2014.

[117] W. Shang, S. Cong, and Y. Ge. Coordination motion control in the task space for parallel manipulators with actuation redundancy. *IEEE Transactions on Automation Science and Engineering*, vol. 10, no. 3, pp. 665–673, 2013.

[118] D. Zhao and Q. Zhu. Position synchronised control of multiple robotic manipulators based on integral sliding mode. *International Journal of Systems Science*, vol. 45, no. January 2014, pp. 556–570, 2014.

[119] L. Chao, Z. Dongya, and X. Xianbo. Force synchronization of multiple robot manipulators: A first study. *Proceedings of the 33rd Chinese Control Conference*, pp. 2212–2217, 2014.

[120] D. Zhao, S. Li, and Q. Zhu. Adaptive synchronised tracking control for multiple robotic manipulators with uncertain kinematics and dynamics. *International Journal of Systems Science*, vol. 7721, no. May 2014, pp. 1–14, 2014.

[121] H. Dou and S. Wang. Robust adaptive motion/force control for motion synchronization of multiple uncertain two-link manipulators. *Mechanism and Machine Theory*, vol. 67, pp. 77–93, 2013.

[122] H. Dou and S. Wang. A boundary control for motion synchronization of a two-manipulator system with a flexible beam. *Automatica*, vol. 50, no. 12, pp. 3088–3099, 2014.

[123] A. Rodriguez-Angeles and H. Nijmeijer. Mutual synchronization of robots via estimated state feedback: A cooperative approach. *IEEE Transactions on Control Systems Technology*, vol. 12, no. 4, pp. 542–554, 2004.

[124] Q. Cao, S. Li, D. Zhao, and Z. Wang. Finite-time motion/force control for motion synchronization of multiple manipulators. *Proceedings of the 33rd Chinese Control Conference, CCC 2014*, pp. 2121–2126, 2014.

[125] R. Cui and W. Yan. Mutual synchronization of multiple robot manipulators with unknown dynamics. *Journal of Intelligent and Robotic Systems: Theory and Applications*, vol. 68, no. 2, pp. 105–119, 2012.

[126] L. D. Khoa, D. Q. Truong, and K. K. Ahn. Synchronization controller for a 3-R planar parallel pneumatic artificial muscle (PAM) robot using modified ANFIS algorithm. *Mechatronics*, vol. 23, no. 4, pp. 462–479, 2013.

Chapter 3

Dynamic Modelling of a Two-link Flexible Manipulator

In this chapter, the mathematical models of a TLFM are developed. An industry prototype of a TLFM is available in the Advanced Control Systems Laboratory, National Institute of Technology, Silchar. These models are developed for designing different robust controllers in the subsequent chapters. This chapter presents the introduction, the derivation of two types of dynamic modelling of a TLFM and the validation of the theoretical model in free and forced evolution. The results and discussion are presented the end of the chapter.

3.1 Introduction

Advancements in manipulator research are categorized into two main parts: rigid manipulators and flexible manipulators (FMs). The current research is more inclined towards FMs because of their several advantages over rigid manipulators. Some significant advantages of FMs include their light weight nature, low energy consumption, compact size, increased workspace, portability, and cost effectiveness. Some of the limitations of flexible manipulators include:

1. Control complexity

- nonminimum phase system [1],
- under actuation problem [2],
- noncollocation [3].

2. Uncertainties

- truncation of flexible modes [4],
- control spillover [5],
- observation spillover [5],
- eigenvalue problem [6].

The main reasons for the above complexities are the choices of a dynamical model [7], and the required structure and operation of FMs [8]. The dynamical model of a flexible manipulator depends on the modelling method. In the last three decades, many methods have been developed for modelling FMs. Generally, three methods are used. These are finite element method (FEM) [9], assumed modes method (AMM) [10] and lumped parameters method [11]. The most widely used method for modelling of FMs is the AMM. It has several advantages like computational efficiency and flexibility in the choice of proper boundary conditions.

3.2 Dynamic model of a two-link flexible manipulator

For improving the performance of a flexible manipulator, the first step is to obtain a reasonably accurate dynamic model which reflects the flexible nature of the links. The mathematical structure of a flexible manipulator is derived from the energy principles. The characteristics of flexible modelling links are:

- They have a rigid as well as a flexible body which is a distributed parameter system [12].
- A manipulator can have a single or multiple links. During operation, its configuration changes and the analysis becomes very complicated.

Physical limitations of a flexible manipulator include:

1. Application of contact forces or torque only at the joints.
2. Finite number of sensors with finite bandwidth are to be used.

In mathematical modelling, if the flexible nature is not taken into account, two kinds of errors may occur. These are the error in the torque requirements of the motors and inaccuracies in positioning the end-effectors. In order to have a precise end-effector positioning, there should ideally be little or no vibration at all. Therefore, to attain precise accuracy field accurate.

In a rigid manipulator, the kinetic and potential energies are stored by virtue of its inertia and gravitational field's position, respectively. But, in a flexible manipulator, the potential energy is stored by the virtue of its flexible links, drives and joints. Joints when modeled as springs can store only potential energy because of their concentrated compliance [13].

Shafts or belts are the drive components and due to their low inertia, they have less kinetic energy. Links are affected by torsion, compression and bending. A torsion stores less kinetic energy and greater potential energy. A compression stores little potential energy because of high stiffness. Bending stores potential energy and kinetic energy because of the deflection rates of a link [13]. The distributed nature of the flexible manipulator is considered for a good modelling.

To include the bending of a flexible link, Euler-Bernoulli equation is often used which neglects rotary inertia and shearing effect. But in [14], both the effects are being incorporated where the beam is generally short relative to its diameter.

The manipulator dynamics are described by a partial differential equation which poses infinite dimensions. Due to the infinite dimensionality of a FM, the dynamic equation is compressed to a finite dimensional model to reduce the constraints on the controller design. The finite element method (FEM) or AMM or LPM is used for the truncation of the FM model.

The modelling scheme of a TLFM using virtual rigid links and passive joints is shown in [15]. The equation of motion of a 2-DOF TLFM is derived and validated by simulation. The parameters of the model are identified by measured data of the real link.

The following assumptions are made to simplify the dynamics of a TLFM for the development of its dynamic model.

- The motion is considered to be in the horizontal plane, focusing on the deflection of the flexible link manipulator.
- The thickness of the beam is negligible as compared to its length. Therefore, transverse shear and rotary inertia effects are negligible.
- Constant cross-sectional area and uniform material properties with constant mass density and Young's modulus are considered.
- The deformation of the links is considered only in the horizontal direction.

The subsequent subsections present brief notes on two widely used modelling methods (lumped parameters and assumed modes methods).

3.2.1 Lumped parameters method

The lumped parameters method models the system as a lump of masses and massless springs. This method is the simplest among the three modelling methods, but generally it doesn't give accurate results.

In Fig. 3.1, (X_0, Y_0) represents the generalised coordinate frame. $(\hat{X}_i, \hat{Y}_i)$ gives the rigid body moving frame and represents the flexible body moving frame associated with the ith link. W_{hi} and J_{hi} are the mass and inertia of the ith hub, respectively. τ_i represents the actuated torque. $\xi_i(x_i, t)$

Table 3.1: The parameters of the master and slave TLFM [16].

Parameter	Link-1	Link-2
Length of links	$L_1 = 0.202\ m$	$L_2 = 0.201\ m$
Mass of links	m_1=0.15268 kg	m_2=0.0535 kg
Elasticity	2.0684x10$^{11}N/m^2$	2.0684x10$^{11}N/m^2$
Resistance of armature	$R_{m1} = 11.5\ \Omega$	$R_{m2} = 2.32\ \Omega$
Equivalent M.I. at load	$J_{eq1} = 0.17043\ kgm^2$	$J_{eq2} = 0.0064387\ kgm^2$
M.I. of the link	$J_{L1} = 0.002035\ kgm^2$	$J_{L2} = 0.0007204\ kgm^2$
Coefficient of viscous damping	$B_{eq1} = 4\ Nms/rad$	$B_{eq2} = 1.5\ Nms/rad$
Efficiency of gear box	$\eta_{g1} = 0.85$	$\eta_{g2} = 0.9$
Efficiency of motor	$\eta_{m1} = 0.85$	$\eta_{m2} = 0.85$
Back e.m.f. constants	$k_{m1} = 0.119\ V/rad$	$k_{m2} = 0.0234\ V/rad$
Gear ratios	$k_{g1} = 100$	$k_{g2} = 50$
Motor torque constants	$k_{t1} = 0.119\ Nm/A$	$k_{t2} = 0.0234\ Nm/A$
Stiffness of link	$k_{s1} = 22\ Nm/rad$	$k_{s2} = 2.5\ Nm/rad$
Maximum Rotation	$(+/-90, +/-90)deg.$	$(+/-90, +/-90)deg.$

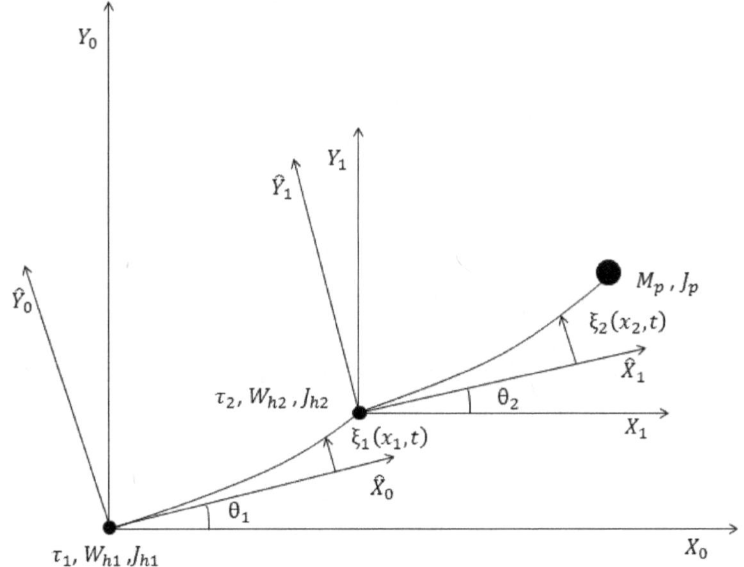

Figure 3.1: A representation of a planar two-link flexible manipulator.

represents the elastic deflection of the ith link. M_p and J_p are the mass and inertia of the payload, respectively, attached at the end of the final link. θ_i is the ith joint angle of the ith link.

The dynamic equations of a TLFM are developed using Euler-Lagrange's equation. The energies of the system, i.e., kinetic energy (K) and potential energy (U) are calculated. K_{Li}, K_{hi} and K_p are the kinetic energies of the ith link, i^{th} hub and payload, respectively, which give the total kinetic energy of the manipulator as given in (3.1)

$$\begin{aligned} K_{total} &= K_{L1} + K_{L2} + K_{h1} + K_{h2} + K_p \\ &= \frac{1}{2}J_{L1}(\dot{\theta}_1 + \dot{\xi}_1)^2 + \frac{1}{2}J_{L2}(\dot{\theta}_2 + \dot{\xi}_2)^2 \\ &+ \frac{1}{2}J_{eq1}\dot{\theta}_1^2 + \frac{1}{2}J_{eq2}\dot{\theta}_2^2 + \frac{1}{2}J_p(\dot{\theta}_2 + \dot{\xi}_2)^2 \end{aligned} \tag{3.1}$$

$$U_{total} = \frac{1}{2}k_{s1}\xi_1^2 + \frac{1}{2}k_{s2}\xi_2^2 \tag{3.2}$$

The Euler-Lagrange equations are,

$$\frac{d}{dt}(\frac{\delta Q}{\delta \dot{\theta}}) - (\frac{\delta Q}{\delta \theta}) = P - B_{eq}\dot{\theta} \tag{3.3}$$

$$\frac{d}{dt}(\frac{\delta Q}{\delta \dot{\xi}}) - (\frac{\delta Q}{\delta \xi}) = 0 \tag{3.4}$$

where,

$$\begin{cases} P = \frac{\eta_m \eta_g k_t k_g (u - k_g k_m \dot{\theta})}{R_m} \\ Q = K_{total} - U_{total} \end{cases} \tag{3.5}$$

We have,

$$\begin{cases} (\frac{\delta Q}{\delta \dot{\theta}_1}) = J_{eq1}\dot{\theta}_1 + J_{L1}(\dot{\theta}_1 + \dot{\xi}_1) \\ \frac{d}{dt}(\frac{\delta Q}{\delta \dot{\theta}_1}) = J_{eq1}\dot{\theta}_1 + J_{L1}(\ddot{\theta}_1 + \ddot{\xi}_1) \end{cases} \tag{3.6}$$

$$\begin{cases} (\frac{\delta Q}{\delta \dot{\xi}_1}) = J_{L1}(\dot{\theta}_1 + \dot{\xi}_1) \\ \frac{d}{dt}(\frac{\delta Q}{\delta \dot{\xi}_1}) = J_{L1}(\ddot{\theta}_1 + \ddot{\xi}_1) \end{cases} \tag{3.7}$$

$$\frac{\delta Q}{\delta \xi_1} = -k_{s1}\xi_1 \tag{3.8}$$

Hence,

$$J_{eq1}\ddot{\theta}_1 + J_{L1}(\ddot{\theta}_1 + \ddot{\xi}_1) = P - B_{eq1}\dot{\theta}_1 \tag{3.9}$$

$$J_{L1}(\ddot{\theta}_1 + \ddot{\xi}_1) + k_{s1}\xi_1 = 0 \tag{3.10}$$

where,

$$\begin{cases} P_1 = [\eta_{m1}\eta_{g1}k_{t1}k_{g1}(u - k_{g1}k_{m1}\dot{\theta}_1)]/R_{m1} \\ P_2 = [\eta_{m2}\eta_{g2}k_{t2}k_{g2}(u - k_{g2}k_{m2}\dot{\theta}_2)]/R_{m2} \end{cases} \tag{3.11}$$

From (3.10), we get,

$$J_{L1}(\ddot{\theta}_1 + \ddot{\xi}_1) = -k_{s1}\xi_1 \tag{3.12}$$

From (3.9) and (3.13), we get,

$$J_{eq1}\ddot{\theta}_1 = P_1 - B_{eq1}\dot{\theta}_1 + k_{s1}\xi_1 \tag{3.13}$$

Putting (3.11) in (3.13)

$$\ddot{\theta}_1 = \frac{k_{s1}\xi_1}{J_{eq1}} + \frac{\eta_{m1}\eta_{g1}k_{t1}k_{g1}u}{J_{eq1}R_{m1}} - (\frac{\eta_{m1}\eta_{g1}k_{t1}k_{g1}^2 + B_{eq1}R_{m1}}{J_{eq1}R_{m1}})\dot{\theta}_1 \tag{3.14}$$

Putting the value of (3.14) in (3.10)

$$J_{L1}\ddot{\xi}_1 = -k_{s1}\xi_1 - J_{L1}\ddot{\theta}_1 \tag{3.15}$$

$$\ddot{\xi}_1 = -k_{s1}\xi_1[\frac{1}{J_{L1}} + \frac{1}{J_{eq1}}] - \frac{\eta_{m1}\eta_{g1}k_{t1}k_{g1}}{J_{eq1}R_{m1}}u + [\frac{\eta_{m1}\eta_{g1}k_{t1}k_{g1}^2 + B_{eq1}R_{m1}}{J_{eq1}R_{m1}}]\dot{\theta}_1 \tag{3.16}$$

Similarly,

$$\ddot{\theta}_2 = \frac{k_{s2}\xi_2}{J_{eq2}} + \frac{\eta_{m2}\eta_{g2}k_{t2}k_{g2}u}{J_{eq2}R_{m2}} - (\frac{\eta_{m2}\eta_{g2}k_{t2}k_{g2}^2 + B_{eq2}R_{m2}}{J_{eq2}R_{m2}})\dot{\theta}_2 \tag{3.17}$$

$$\ddot{\xi}_2 = -k_{s2}\xi_2[\frac{1}{J_{L2}} + \frac{1}{J_{eq2}}] - \frac{\eta_{m2}\eta_{g2}k_{t2}k_{g2}}{J_{eq2}R_{m2}}u + [\frac{\eta_{m2}\eta_{g2}k_{t2}k_{g2}^2 + B_{eq2}R_{m2}}{J_{eq2}R_{m2}}]\dot{\theta}_2 \tag{3.18}$$

$$\begin{cases} p_1 = \frac{\eta_{m1}\eta_{g1}k_{t1}k_{g1}^2k_{m1}+B_{eq1}R_{m1}}{R_{m1}J_{eq1}} \\ p_2 = \frac{k_{s1}}{J_{eq1}} \\ p_3 = \frac{\eta_{m1}\eta_{g1}k_{t1}k_{g1}}{R_{m1}J_{eq1}} \\ p_4 = \frac{\eta_{m2}\eta_{g2}k_{t2}k_{g2}^2k_{m2}+B_{eq2}R_{m2}}{R_{m2}J_{eq2}} \\ p_5 = \frac{k_{s2}}{J_{eq2}} \\ p_6 = \frac{\eta_{m2}\eta_{g2}k_{t2}k_{g2}}{R_{m2}J_{eq2}} \\ p_7 = k_{s1}[\frac{1}{J_{L1}} + \frac{1}{J_{eq1}}] \\ p_8 = k_{s2}[\frac{1}{J_{L2}+J_p} + \frac{1}{J_{eq2}}] \end{cases} \tag{3.19}$$

where $p_{i=1,\ldots,8}$ are the parameters of the system as in (3.19). Descriptions of the variables, parameters and their values are given as in Table 3.1. Finally, the mathematical model of the TLFM in LPM can be written as

$$\begin{pmatrix} \dot{\theta}_1 \\ \dot{\theta}_2 \\ \dot{\xi}_1 \\ \dot{\xi}_2 \\ \ddot{\theta}_1 \\ \ddot{\theta}_2 \\ \ddot{\xi}_1 \\ \ddot{\xi}_2 \end{pmatrix} + \begin{pmatrix} 0 & 0 & 0 & 0 & 1 & 0 & 0 & 0 \\ 0 & 0 & 0 & 0 & 0 & 1 & 0 & 0 \\ 0 & 0 & 0 & 0 & 0 & 0 & 1 & 0 \\ 0 & 0 & 0 & 0 & 0 & 0 & 0 & 1 \\ 0 & 0 & p_2 & 0 & -p_1 & 0 & 0 & 0 \\ 0 & 0 & 0 & p_5 & 0 & -p_4 & 0 & 0 \\ 0 & 0 & -p_7 & 0 & p_1 & 0 & 0 & 0 \\ 0 & 0 & 0 & -p_8 & 0 & p_4 & 0 & 0 \end{pmatrix} \begin{pmatrix} \theta_1 \\ \theta_2 \\ \xi_1 \\ \xi_2 \\ \dot{\theta}_1 \\ \dot{\theta}_2 \\ \dot{\xi}_1 \\ \dot{\xi}_2 \end{pmatrix} + \begin{pmatrix} 0 \\ 0 \\ 0 \\ 0 \\ p_3 \\ p_6 \\ -p_3 \\ -p_6 \end{pmatrix} u \tag{3.20}$$

(3.20) can be then segregated into two stages as in (3.21) and (3.22)

$$\begin{pmatrix} \dot{\theta}_1 \\ \dot{\xi}_1 \\ \ddot{\theta}_1 \\ \ddot{\xi}_1 \end{pmatrix} = \begin{pmatrix} 0 & 0 & 1 & 0 \\ 0 & 0 & 0 & 1 \\ 0 & p_2 & -p_1 & 0 \\ 0 & -p_7 & p_1 & 0 \end{pmatrix} \begin{pmatrix} \theta_1 \\ \xi_1 \\ \dot{\theta}_1 \\ \dot{\xi}_1 \end{pmatrix} + \begin{pmatrix} 0 \\ 0 \\ p_3 \\ -p_3 \end{pmatrix} u_1 \tag{3.21}$$

$$\begin{pmatrix} \dot{\theta}_2 \\ \dot{\xi}_2 \\ \ddot{\theta}_2 \\ \ddot{\xi}_2 \end{pmatrix} = \begin{pmatrix} 0 & 0 & 1 & 0 \\ 0 & 0 & 0 & 1 \\ 0 & p_5 & -p_4 & 0 \\ 0 & -p_8 & p_4 & 0 \end{pmatrix} \begin{pmatrix} \theta_1 \\ \xi_1 \\ \dot{\theta}_2 \\ \dot{\xi}_2 \end{pmatrix} + \begin{pmatrix} 0 \\ 0 \\ p_6 \\ -p_6 \end{pmatrix} u_2 \tag{3.22}$$

3.2.2 Assumed modes method

References [7], [8] developed the motion equations that include the translational motion of elastic member of the standard arm. In the assumed modes method an infinite dimensional model of a system is truncated to a finite dimensional model series in terms of a finite varying mode amplitude and the spatial eigen mode function [17], [18]. The boundary conditions of links and their mode eigen functions can be chosen using the method given in [19]. A schematic diagram of a TLFM is shown in Fig. 3.1.

In reference [20], AMM modelling of a TLFM was studied. The two-link flexible arm dynamic characteristic, modelling and vibration control with payload, gravity and the coupled vibration between the two arms are considered. The proper boundary conditions of Euler beam equations are also considered, where the governing equations are derived using the modal analysis method matrix.

In the AMM method, the dynamic model of the deflection of a flexible link is discussed by a set of the link vibration modes apart from its natural modes. Hence, the flexible deflection is decomposed into a combination of mode eigenfunctions $\Phi_{ik}(x_i)$ which are also called mode shapes and time dependent generalised coordinates $\delta_{ik}(t)$ as described in (3.23).

$$\xi_i(x_i,t) = \sum_{k=0}^{r_i} \Phi_{ik}(x_i)\delta_{ik}(t) \tag{3.23}$$

where $\xi_i(x_i,t)$ is the deflection of the ith link at $x_i(0 \leq x_i \leq l_i)$, l_i is the length of the ith link and r_i is the number of modes used to describe the deflection of link i. $\Phi_{ik}(x_i)$ is the kth mode shape (spatial coordinate) function of link i. $\delta_{ik}(t)$ is the kth modal coordinate (time coordinate) of the ith link. When $r_i = 0$, for the ith link, it is called the zeroth mode, which gives characteristics similar to that of a rigid manipulator.

There are various ways of choosing the boundary conditions for the AMM method. Selection of suitable boundary conditions of the AMM can be critical for FMs to fit into an application. Usually, boundary conditions are selected based on the set nearest to the natural modes of the system. However, natural modes depend on several factors of the system including the size of the payload mass and the hub inertia. Hence, the choice of appropriate boundary conditions is very important as it gives better results and further consequences. The final choice of boundary conditions requires an evaluation based on the actual manipulator structure and its payload range together with its natural modes [21].

The flexible deflection $\xi_i(x_i,t)$ can be described as in (3.24)

$$\begin{cases} \xi_i(x_i,t) = \sum_{k=1}^{r_i} \Phi_{ik}(x_i)\delta_{ik}(t) \\ \delta_{ik}(t) = exp(j\omega_{ik}t) \end{cases} \tag{3.24}$$

where ω_{ik} is the kth natural angular frequency for the ith link and its value is found by simulation consequently. The flexible manipulator links are modelled as Euler-Bernoulli beams where ρ_i denotes mass per unit length. $(EI)_i$ is the constant flexural rigidity where $\xi_i(x_i,t)$ is the flexible deflection satisfying the partial differential equation in (3.25) [22].

$$(EI)_i \frac{\partial^4 \xi_i(x_i,t)}{\partial x_i^4} + \rho_i \frac{\partial^2 \xi_i(x_i,t)}{\partial t^2} = 0, \quad i = 1,...,n \tag{3.25}$$

$$(EI)_i \frac{\partial^4 \xi_i(x_i,t)}{\partial x_i^4} = -\rho_i \frac{\partial^2 \xi_i(x_i,t)}{\partial t^2} = 0 \tag{3.26}$$

$$\frac{\frac{\partial^4 \xi_i(x_i,t)}{\partial x_i^4}}{\frac{\partial^2 \xi_i(x_i,t)}{\partial t^2}} = \frac{-\rho}{EI} \quad [\because \xi_i(x_i,t) = \Phi_{ik}(x_i)\delta_{ik}(t)] \tag{3.27}$$

$$\frac{\partial_{ik}(t)\frac{\partial^4 \Phi_{ik}(x_i,t)}{\partial x_i^4}}{\Phi_{ik}(x)\frac{\partial^2 \delta_{ik}(t)}{\partial t^2}} = \frac{-\rho}{EI} \tag{3.28}$$

$$\frac{\partial^4 \Phi_{ik}(x)/\partial x^4}{\partial^2 \delta_{ik}(t)/\partial t^2} = \frac{-\rho}{EI}\frac{\Phi_{ik}(x)}{\delta_{ik}(t)} \tag{3.29}$$

$$\frac{\partial^4 \Phi_{ik}(x)/\partial x^4}{\partial^2 \delta_{ik}(t)/\partial t^2} = \frac{-\beta_{ik}^4}{-\omega_{ik}^2}\frac{\Phi_{ik}(x)}{\delta_{ik}(t)} \tag{3.30}$$

Here, we can segregate (3.32) into two equations

$$\frac{\partial^4 \Phi_{ik}(x)}{\partial x^4} = \beta_{ik}^4 \Phi_{ik}(x) \tag{3.31}$$

$$\begin{cases} \frac{\partial^4 \Phi_{ik}(x)}{\partial x^4} = \beta_{ik}^4 \Phi_{ik}(x) \\ \frac{\partial^4 \Phi_{ik}(x)}{\partial x^4} - \beta_{ik}^4 \Phi_{ik}(x) = 0 \end{cases} \tag{3.32}$$

$$\begin{cases} \frac{\partial^2 \delta_{ik}(t)}{\partial t^2} = -\omega_{ik}^2 \delta_{ik}(t) \\ \frac{\partial^2 \delta_{ik}(t)}{\partial t^2} + \omega_{ik}^2 \delta_{ik}(t) = 0 \end{cases} \tag{3.33}$$

The solution of (3.33) is,

$$\delta(t) = C_1 e^{i\omega_{ik}t} + C_2 e^{-i\omega_{ik}t} \tag{3.34}$$

where the second term in (3.34) can be neglected due to the decaying function and $C_1 = 1$ (assumed). Hence,

$$\delta(t) = exp(i\omega_{ik}t) \tag{3.35}$$

The solution of (3.32) can is,

$$\Phi(x) = C_1 \cos \beta_{ik}x + C_2 \sin \beta_{ik}x + C_3 \cosh \beta_{ik}x + C_4 \sinh \beta_{ik}x \tag{3.36}$$

C_1, C_2, C_3, C_4 are the unknown coefficients. The boundary conditions are given in (3.37) as each link is considered to be clamped at the base and free at the end. It is assumed that the lightweight link inertia is small compared to the hub inertia and then the contrained mode shapes can be used [11].

$$\xi_i(0,t) = 0, \quad \xi_i'(0,t) = 0, \quad i = 1, ..., n \tag{3.37}$$

In regard to the remaining boundary conditions, the link end free of dynamic constraints is assumed due to the difficulty of accounting for time-varying or unknown masses and inertias. However, it is better to consider mass boundary conditions representing balance moments and shearing force moments [23],

Bending moment:

$$\begin{cases} (EI)_i \frac{\partial^2 \xi_i(x_i,t)}{\partial x_i^2} |_{x_i=l_i} = & - J_{Ti}\frac{d^2}{dt^t}(\frac{\partial^2 \xi_i(x_i,t)}{\partial x_i} |_{x_i=l_i}) \\ & - (M_R)_i \frac{d^2}{dt^2}(\xi_i(x_i,t) |_{x_i=l_i}) \quad i = 1, ..., n \end{cases} \tag{3.38}$$

Shearing moment:

$$\begin{cases} (EI)_i \dfrac{\partial^3 \xi_i(x_i,t)}{\partial x_i^3} \mid_{x_i=l_i} = M_{Ti} \dfrac{d^2}{dt^2}(\xi_i(x_i,t) \mid_{x_i=l_i}) \\ \qquad (M_R)_i \dfrac{d^2}{dt^2}(\dfrac{\partial^2 \xi_i(x_i,t)}{\partial x_i} \mid_{x_i=l_i}) \quad i=1,...,n \end{cases} \tag{3.39}$$

where M_{Ti}, J_{Ti} are the actual mass and moment of inertia, respectively, at the end of the ith link. $(M_R)_i$ is the total contribution of the masses non-collocated at the end of the link i, weighted by the relative distance from axis Y_i (shearing axis at the end of link i). Somehow, these contributions are often not included in the mode shape analyses [23].

The total mass and the total moment of inertia of a TLFM are given as

For link-1,

$$\begin{cases} M_{T1} = m_2 + m_{h2} + M_p \\ J_{T1} = J_{02} + J_{h2} + J_p + M_p l_2^2 \end{cases} \tag{3.40}$$

For link-2,

$$\begin{cases} B_{T2} = M_p \\ J_{T2} = J_p \end{cases} \tag{3.41}$$

where m_2 and m_{h2} are the masses of link-2 and hub-2, respectively. The mass of the payload attached at the end of link-2 is denoted as M_p. J_{02} is the inertia of the link-2 about the joint-2 axis. J_{h2} and J_p are the moment of inertia of the hub-2 and payload, respectively.

Applying the boundary conditions and $\zeta_{ik}^4 = \frac{\omega_{ik}^2 \rho_i}{(EI)_i}$, in (3.37) and using (3.36), we get

$$C_{3,ik} = -C_{1,ik}, \quad C_{4,ik} = -C_{2,ik} \tag{3.42}$$

Giving the homogeneous system of equations of the form

$$\left(\begin{array}{c} F(\zeta_{ik}) \end{array} \right) \left(\begin{array}{c} C_{1,ik} \\ C_{2,ik} \end{array} \right) = 0 \tag{3.43}$$

From (3.42) and (3.43), we get

$$\begin{aligned} \Phi_{ik}(x_i) = C_{1,ik}[&\sin(\beta_{ik}x_i) - \sinh(\beta_{ik}x_i) \\ &+ \varrho\{\cos(\beta_{ik}x_i) - \cosh(\beta_{ik}x_i)\}] \end{aligned} \tag{3.44}$$

where $D_{1,ik} = 0.5$ and $\varrho = \frac{D_{2,ik}}{D_{1,ik}}$ are simulated for each mode. By setting (3.43) to zero, the frequency of the equation is obtained which further gives a transcendental equation.

In the following, we continue only with the nominal payload condition. On the basis of equations introduced in the previous section, the Lagrangian $'L'$ becomes a set of generalised coordinates $'q_i(t)'$. The dynamic model is obtained by satisfying the Lagrangian-Euler equations.

$$\frac{d}{dt}\frac{\delta L}{\delta \dot{q}_i} - \frac{\delta L}{\delta q_i} = F_i, \quad i = 1, ..., N \tag{3.45}$$

where F_i are the generalised forces acting on q_i. Based on the Lagrange's assumed modes method (AMM), the dynamic equations of a TLFM [24] can be written as

$$B(\theta_i, \upsilon_{ik})\begin{pmatrix} \ddot{\theta}_i \\ \ddot{\upsilon}_{ik} \end{pmatrix} + h(\theta, \dot{\theta}, \upsilon_{ik}, \dot{\upsilon}_{ik}) + K\begin{pmatrix} \theta_i \\ \upsilon_{ik} \end{pmatrix} + D\begin{pmatrix} \dot{\theta}_i \\ \dot{\upsilon}_{ik} \end{pmatrix} = b\tau \tag{3.46}$$

Considering the coordinate notation $q = [\theta_1,\ \theta_2,\ \upsilon_{11},\ \upsilon_{12},\ \upsilon_{21},\ \upsilon_{22}]^T$, (6.19) can be simplified as

$$B(q)\ddot{q} + h(q, \dot{q}) + Kq + D\dot{q} = b(\tau + \tau_d) \tag{3.47}$$

It is assumed that the robotic manipulators have uncertainties where $B(q) = B_0(q) + \Delta B(q) \in R^{6X6}$ is the inertia matrix, $h(q, \dot{q}) = h_0(q, \dot{q}) + \Delta h(q, \dot{q}) \in R^{6X1}$ is the centrifugal and coriolis force vector, $K = K_0 + \Delta K$ is the positive definite stiffness matrix $\in R^{6X6}$, $D = D_0 + \Delta D$ is the positive definite damping matrix $\in R^{6X6}$, $\tau \in R^{2X1}$ is the joint torque input, $\tau_d \in R^{2X1}$ is the disturbance torque input and $b \in R^{6X2}$ is the input weighting matrix. Here, $B_0(q), h_0(q, \dot{q}), K_0, D_0$ and $\Delta B(q), \Delta h(q, \dot{q}), \Delta K, \Delta D$ represent the nominal and bounded perturbations, respectively. Then, (6.19) can be represented as

$$B_0(q)\ddot{q} + h_0(q, \dot{q}) + K_0 q + D_0\dot{q} = b\tau + b\tau_d + P(q, \dot{q}, \ddot{q}) \tag{3.48}$$

where,

$$P(q, \dot{q}, \ddot{q}) = -\Delta B(q)\ddot{q} - \Delta h(q, \dot{q}) - \Delta K q - \Delta D\dot{q}$$

is the collective system uncertainty and is bounded as in (3.49) [25] as

$$\|P(q, \dot{q}, \ddot{q})\| < \gamma_0 + \gamma_1\|q\| + \gamma_2\|\dot{q}\|^2 \leq \Delta\gamma(q, t) \tag{3.49}$$

where $\gamma_0, \gamma_1, \gamma_2$ are positive constants.

It is also assumed that the manipulator dynamics satisfies the following conditions [26]

$$\|B_0(q)\| < \mu_0 \tag{3.50}$$

$$\|\tau(t)\| < \mu_1 + \mu_2\|q\| + \mu_3\|\dot{q}\|^2 \tag{3.51}$$

where $\mu_0, \mu_1, \mu_2, \mu_3$ are positive constants.

3.3 Results and discussion

A set of numerical simulations have been performed for validating the theoretical model. The simulations are carried out using ode-45 solver in a MATLAB® simulation environment. The mode shapes for no payload, nominal payload and maximum payload conditions are shown in the subsequent figures. The natural frequencies are calculated by $f_{ik} = \omega_{ik}/2\pi$ and obtained from $(\zeta_{ik})^4 = \frac{(\omega_{ik})^2 \rho_i}{(EI)_i}$ and the values are as follows in (3.52):

$$\begin{cases} \zeta_{11} = 2.6302, & \omega_{11} = 8.5303, & f_{11} = 1.3576 \\ \zeta_{12} = 5.9953, & \omega_{12} = 44.3212, & f_{12} = 7.0539 \\ \zeta_{21} = 3.7567, & \omega_{21} = 48.439, & f_{21} = 7.7093 \\ \zeta_{22} = 7.655, & \omega_{22} = 201.1287, & f_{22} = 32.0106 \end{cases} \tag{3.52}$$

where the stiffness constant K is given as $K = diag\{0, 0, \omega_{11}^2 m_1, \omega_{12}^2 m_1, \omega_{21}^2 m_2, \omega_{22}^2 m_2\}$. Figure 3.2 depicts the values of β for link-1 and link-2 from which the natural frequencies are obtained. Figures 3.3, 3.3, 3.4 show the mode shapes of link-1 and link-2 with the y-axis as the link length. It is also seen that the change in the boundary condition values $M_{L1}, J_{L1}, M_{L2}, J_{L2}$ does not modify the shapes of link-1 and link-2. Further, the parameters involved in the computation are computed as:

$$\begin{cases} \phi_{11,e} = 0.2955, & \phi_{12,e} = 0.2518 \\ \phi_{21,e} = 0.4909, & \phi_{22,e} = 0.2226 \\ \phi\prime_{11,e} = 2.8032, & \phi\prime_{12,e} = -0.3770 \\ \phi\prime_{21,e} = 4.5152, & \phi\prime_{22,e} = -0.6792 \\ v_{11} = 0.0184, & v_{12} = -0.0243 \\ v_{21} = 0.0105, & v_{22} = 0.0077 \\ w_{11} = 0.0028, & w_{12} = 0.0034 \\ w_{21} = 0.0016, & w_{22} = 0.0011 \end{cases} \tag{3.53}$$

3.3.1 Validation of the AMM model under free and forced conditions

In this subsection, validation of the theoretical model is performed in the presence of free and forced evolution. Figure 3.6 shows the initial condition dependence resulting in internal vibrations of the link modes. An initial condition is subjected onto the link-2 $(\delta_{21}(0) = 0.1, \delta_{22}(0) = 0.002)$ in Fig. 3.6 showing the vibration coupling between the links. Figure 3.7 shows the joint tracking where relative drifting is observed. In Fig. 3.8, link-2 is kept at a right angle to link-1 $(\theta_2(0) = \pi/2)$ with an initial condition $(\delta_{11}(0) = 0.1, \delta_{12}(0) = 0.002)$ on link-1. In Fig. 3.9, deflections of link-1 and link-2 are shown where structural damping is added along with the initial condition $(\delta_{21}(0) = 0.1, \delta_{22}(0) = 0.002)$ showing an improvement in the deflection of the links. Figure 3.10 shows the joint motion of link-1 and link-2 under the same initial condition. Figure 3.11 shows the modes deflection of link-1 and link-2 with the same initial condition in addition to link-1 kept at a right angle to link-2. A bang-bang input torque of 0.032 N (symmetrical) is shown in Fig. 3.12. The

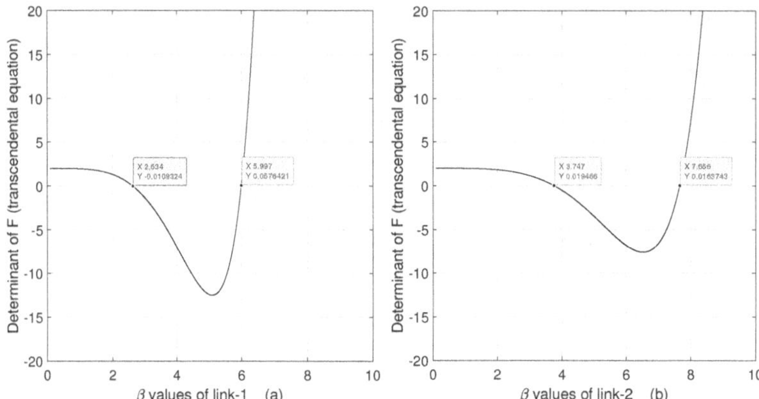

Figure 3.2: β values for link-1 and link-2.

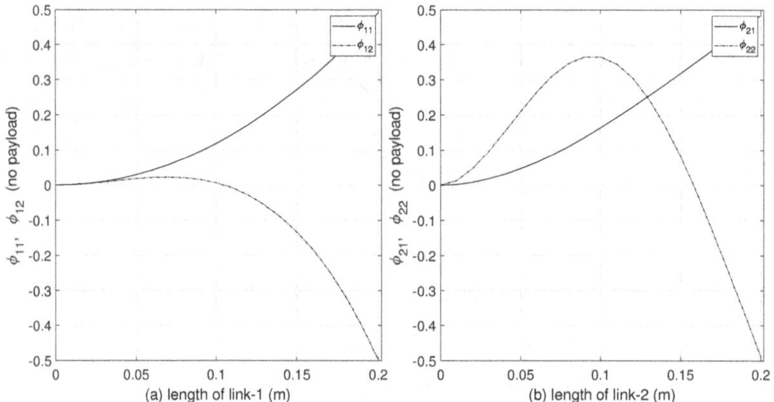

Figure 3.3: Mode shapes of link-1 and link-2 under no payload condition.

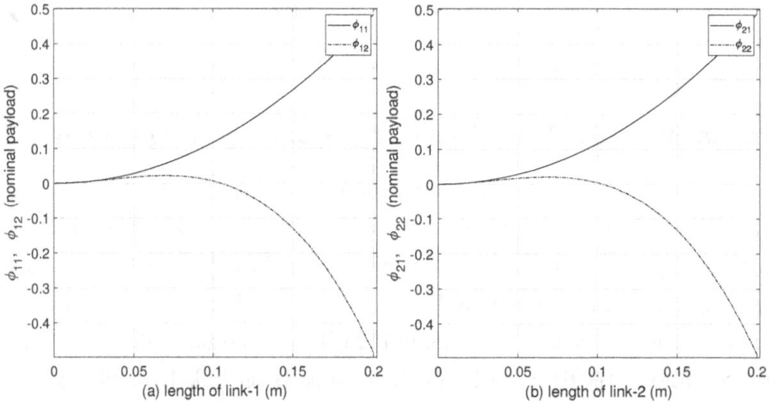

Figure 3.4: Mode shapes of link-1 and link-2 under nominal payload condition.

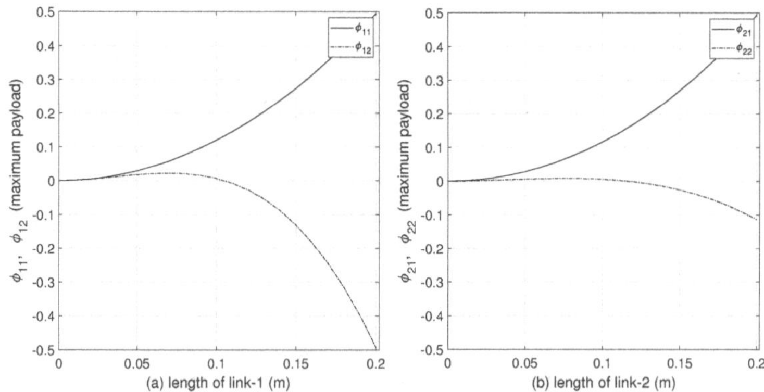

Figure 3.5: Mode shapes of link-1 and link-2 under maximum payload condition.

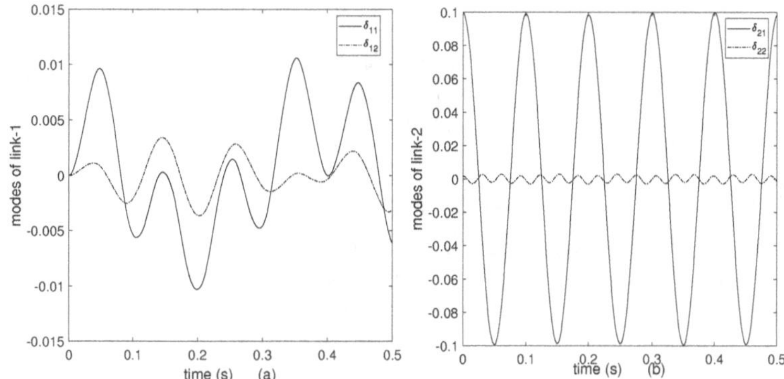

Figure 3.6: Deflection of link-1 and link-2 with the initial conditions $(\theta_1(0) = 0, \theta_2(0) = 0, \delta_{11}(0) = 0, \delta_{12}(0) = 0, \delta_{21}(0) = 0.1, \delta_{22}(0) = 0.002)$.

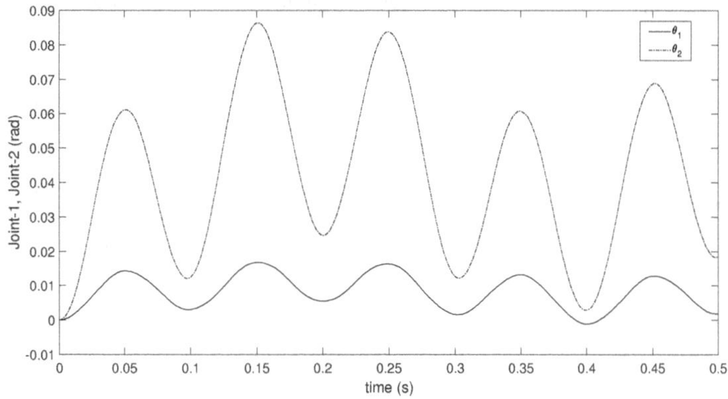

Figure 3.7: Joint motion of link-1 and link-2 with the initial conditions $(\theta_1(0) = 0, \theta_2(0) = 0, \delta_{11}(0) = 0, \delta_{12}(0) = 0, \delta_{21}(0) = 0.1, \delta_{22}(0) = 0.002)$.

bang-bang torque in Fig. 3.12 is applied at both the joints and the vibration of the modes for link-1 and link-2 is shown in Fig. 3.13 with the introduction of damping. These vibrations can be damped and smothered which is shown in Fig. 3.14.

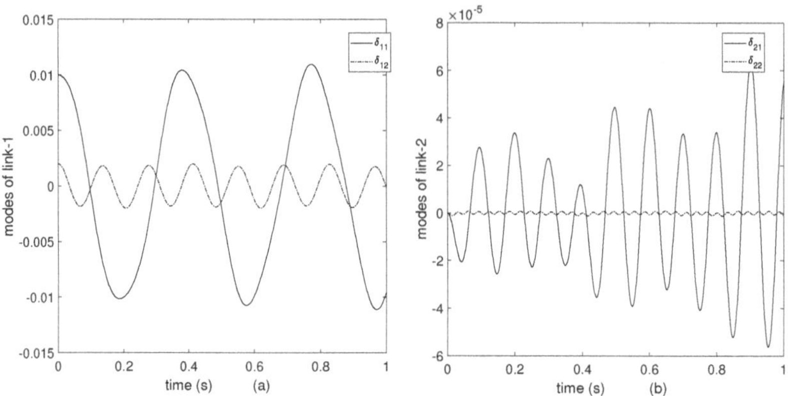

Figure 3.8: Deflection of link-1 and link-2 with the initial conditions $(\theta_1(0) = 0, \theta_2(0) = \pi/2, \delta_{11}(0) = 0.1, \delta_{12}(0) = 0.002, \delta_{21}(0) = 0, \delta_{22}(0) = 0)$.

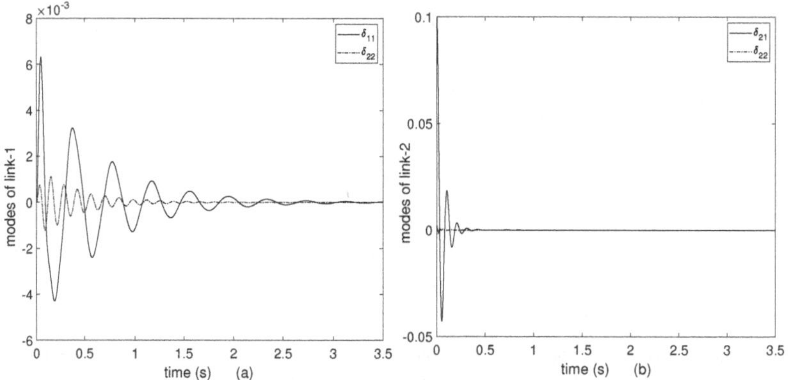

Figure 3.9: Deflection of link-1 and link-2 with the initial conditions $(\theta_1(0) = 0, \theta_2(0) = 0, \delta_{11}(0) = 0, \delta_{12}(0) = 0, \delta_{21}(0) = 0.1, \delta_{22}(0) = 0.002)$ along with damping.

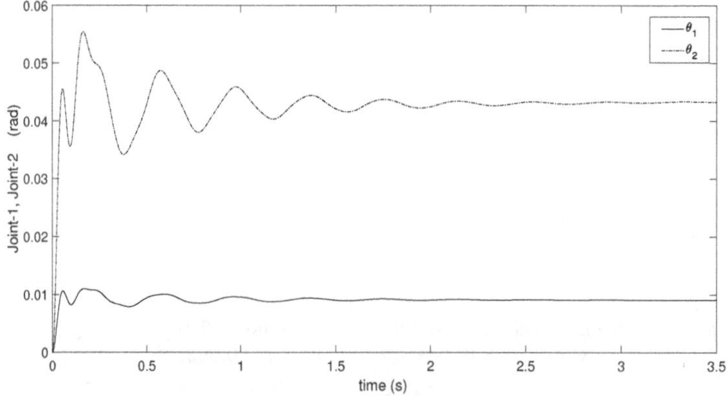

Figure 3.10: Joint motion of link-1 and link-2 with the initial conditions $(\theta_1(0) = 0, \theta_2(0) = 0, \delta_{11}(0) = 0, \delta_{12}(0) = 0, \delta_{21}(0) = 0.1, \delta_{22}(0) = 0.002)$ along with damping.

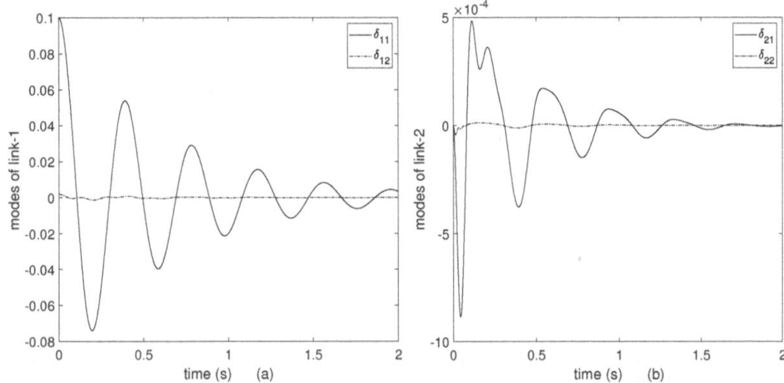

Figure 3.11: Deflection of link-1 and link-2 with the initial conditions $(\theta_1(0) = 0, \theta_2(0) = \pi/2, \delta_{11}(0) = 0, \delta_{12}(0) = 0, \delta_{21}(0) = 0.1, \delta_{22}(0) = 0.002)$ along with damping.

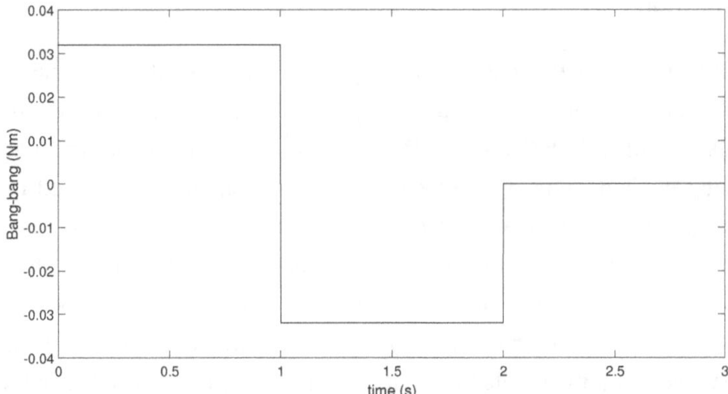

Figure 3.12: A symmetrical bang-bang input torque of 0.032 N.

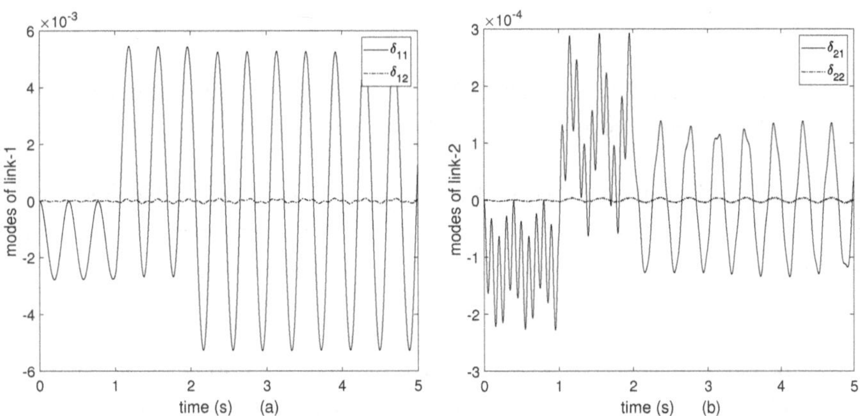

Figure 3.13: Deflection of link-1 and link-2 when the bang-bang input torque is applied with the initial conditions $(\theta_1(0) = 0, \theta_2(0) = 0, \delta_{11}(0) = 0, \delta_{12}(0) = 0, \delta_{21}(0) = 0, \delta_{22}(0) = 0)$.

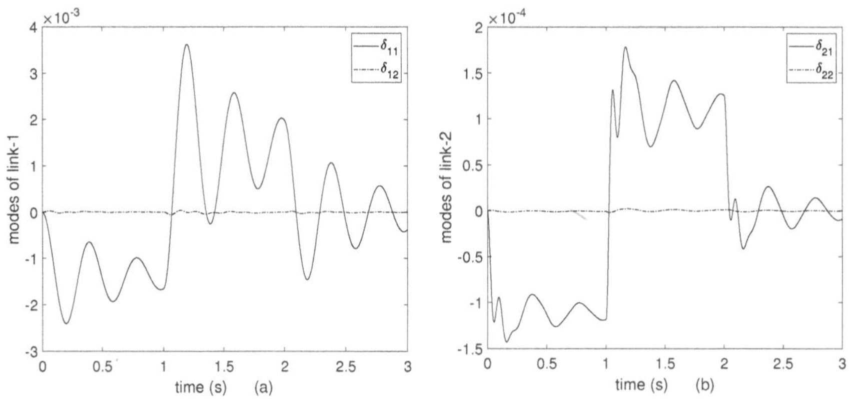

Figure 3.14: Deflection of link-1 and link-2 when the bang bang input torque is applied and damping with the initial conditions $(\theta_1(0) = 0, \theta_2(0) = 0, \delta_{11}(0) = 0, \delta_{12}(0) = 0, \delta_{21}(0) = 0, \delta_{22}(0) = 0)$.

3.4 Chapter summary

This chapter described the dynamic equations of a TLFM using two different methods. The final equations derived will be used in the subsequent chapters for designing different control algorithms. The dynamic equations of a TLFM are obtained by using the Lagrangian approach along with respective modelling methods, namely, lumped parameters method and assumed modes method. Emphasis is given on obtaining accurate equations of motion that display the coupling between the rigid as well as flexible dynamics. The mode shape analysis is also exploited with time-varying mass boundary conditions considering two modes for each link. This theoretical model is tested under free and forced conditions, i.e., by inducing a symmetrical bang-bang torque signal. Additionally the inclusion of passive damping and its beneficial effects are clearly illustrated in the figures. The results demonstrate that the model obtained effectively represents the dynamics of a TLFM.

References

[1] D. Chen and B. Paden. Stable Inversion of Nonlinear Non-minimum Phase Systems. *International Journal of Control*, vol. 64, pp. 81–97, May 1996.

[2] X. Xin. Swing-up Control for a Two-Link Underactuated Robot with a Flexible Elbow Joint: New Results beyond the Passive Elbow Joint. In 50th *IEEE Conference on Decision and Control and European Control Conference (CDC-ECC)*, (Orlando, FL, USA), pp. 2481–2486, December 2011.

[3] V. A. Spector and H. Flashner. Modeling and Design Implications of Noncollocated Control in Flexible Systems. *Journal of Dynamical System Measurement and Control*, vol. 112, pp. 186–193, June 1990.

[4] N. Zhang, Y. Feng, and X. Yu. Optimization of Terminal Sliding Control for Two-link Flexible Manipulators. In 30th *IEEE annual Conference of the Industrial Electronics Society*, vol. 2, (Busan, Korea), pp. 1318–1322, November 2004.

[5] F. Khorrami and S. Jain. Experimental Results on an Inner/Outer Loop Controller for a Two-Link Flexible Manipulator. In *Proc. of the IEEE International Conference on Robotics and Automation*, vol. 1, pp. 742–747, May 1992.

[6] W. J. Book, O. Maizza-Neto, and D. E. Whitney. Feedback Control of Two Beam, Two Joint Systems with Distributed Flexibility. *Journal of Dynamical System Measurement and Control*, vol. 97, pp. 424–431, December 1975.

[7] K. W. Buffinton and T. R. Kane. Dynamics of Beam Moving Over Supports. *International Journal of Solids Structures*, vol. 21, no. 7, pp. 617–643, 1985.

[8] K. W. Buffinton. Dynamics of Elastic Manipulators with Prismatic Joints. *Journal of Dynamic Systems Measurement and Control*, vol. 114, pp. 41–49, March 1992.

[9] B. Jonker. A Finite Element Dynamic Analysis of Flexible Spatial Mechanisms with Flexible Links. *Computational Methods in Applied Mechanics and Engineering*, vol. 76, pp. 17–40, November 1989.

[10] A. S. Morris and A. Madani. Quadratic Optimal Control of a Two-Flexible-Link Robot Manipulator. *Robotica*, vol. 16, pp. 97–108, January 1988.

[11] J. H. Kim and M. Uchiyama. Dynamic Modeling of Two Cooperating Flexible Manipulators. *KSME International Journal*, vol. 4, pp. 188–196, February 2000.

[12] H. Kanoh and H. G. Lee. Vibration Control of One-link Flexible Arm. In *Proc. of* 24th *IEEE Conference on Decision and Control*, vol. 3, (Fort Lauderdale, USA), pp. 1172–1177, December 1985.

[13] S. K. Dwivedy and P. Eberhard. Dynamic Analysis of Flexible Manipulators, a Literature Review. *Mechanism and Machine Theory*, vol. 41, pp. 749–777, July 2006.

[14] S. Timoshenko. *Vibration Problems in Engineering*. John Wiley and Sons, 2nd ed., 1946.

[15] T. Yoshikawa and K. Hosoda. Modeling of Flexible Manipulators Using Virtual Rigid Links and Passive Joints. In *IEEE International Workshop on Intelligent Robots and Systems*, vol. 2, pp. 967–972, 1991.

[16] Quanser. Equation For The Frist (Second) Stage of The 2DOF Serial Flexible Link Robot. In *QUANSER*, 2006.

[17] L. Meirovitch. *Methods of Analytical Dynamics*. McGraw-Hill, 1970.

[18] G. E. Meyer. *Analytical Methods in Conduction Heat Transfer*. McGraw-Hill, 1971.

[19] H. N. Rahimi and M. Nazemizadeh. Dynamic Analysis and Intelligent Control Techniques for Flexible Manipulators: A Review. *Advanced Robotics*, vol. 28, pp. 63–76, January 2014.

[20] T. Fukuda. Flexibility Control of Elastic Robotic Arms. *Journal of Robotic System*, vol. 2, no. 1, pp. 73–88, 1985.

[21] M. H. Korayem, H. N. Rahimi, and M. Nikoobin. Dynamic Analysis, Simulation and Trajectory Planning of Mechanical Mobile Manipulators with Flexible Links and Joints. *Applied Mathematical Modelelling*, vol. 36, no. 29, pp. 32–44, 2012.

[22] B. Subudhi and A. S. Morris. Dynamic Modelling, Simulation and Control of A Manipulator with Flexible Links and Joints. *Rob. Auton. Syst.*, vol. 41, no. 4, pp. 257–270, 2002.

[23] A. DeLuca and B. Siciliano. Closed-form dynamic model of planar multilink lightweight robots, volume = 21, year = 1991. *Trans. Syst. Man Cybern.*, no. 4, pp. 826–839.

[24] S. K. Pradhan and B. Subudhi. Nonlinear Adaptive Model Predictive Controller for A Flexible Manipulator: An Experimental Study. *IEEE Trans. Control Syst. Technol.*, vol. 22, pp. 1754–1768, 2014.

[25] B. K. Yoo and W. C. Ham. Adaptive control of robot manipulator using fuzzy. *IEEE Transactions on Fuzzy Systems*, vol. 8, no. 2, pp. 186–199, 2000.

[26] X. T. Tran and H. J. Kang. Adaptive hybrid high-Order terminal sliding mode control of MIMO uncertain nonlinear systems and its application to robot manipulators. *International Journal of Precision Engineering and Manufacturing*, vol. 16, no. 2, pp. 255–266, 2015.

Chapter 4

Design of Sliding Mode Controllers for a TLFM

In this chapter, focus is on the problem of tip trajectory tracking control while also addressing the suppression of tip deflection in a planar assumed modes modelled two-link flexible manipulator (TLFM). A robust chattering free finite time second-order SMC is designed to address this control problem. Conventional first order sliding surfaces are designed in terms of the tracking error first. Then, the second-order sliding surfaces are designed in terms of the first order SMC. The performance metrics considered are tracking error and time for the suppression of the tip deflection. The simulation results are compared with the existing SMC based techniques for tracking control of a TLFM. Better performances of the proposed controller are visibly reflected in the simulation results in a MATLAB® environment.

The chapter is organized as follows: - Chapter objective - Structure and basics of sliding mode controller - Development of the proposed conventional SMC and second-order sliding mode control for a two-link flexible manipulator - Presentation of results and discussion - Chapter summary.

4.1 Introduction

As discussed in Chapter 2, flexible manipulators prove to be more effective from an application perspective in various fields such as space, industries, defense, hospitals and home appliances in comparison to rigid manipulators [1], [2]. The research in flexible manipulators is predominantly focused two-link flexible manipulators (TLFM) compared to other types of flexible manipulators [3], [4]. Furthermore, the tip trajectory tracking control represents one of the most widely accepted and challenging problems in this domain.

Many control techniques have been reported for the tip trajectory tracking control of TLFMs like adaptive control [1], model predictive control (MPC) [4], sliding mode control (SMC), observer based [5] and fuzzy logic control (FLC) [6]. But, designing an appropriate controller for the tip trajectory tracking control of a TLFM is still a challenging task. A SMC is an efficient and effective control technique compared to other control techniques [7]. The major advantages of the SMC are its robustness to model and external uncertainties, insensitivity to disturbances, fast response, global stability, finite time stability, and more [8]. Various SMC techniques have been reported for the various control problems of TLFMs. The reported papers where SMC techniques are used for TLFMs are categorised in Table 4.1.

It is seen that very few papers are available for the tip trajectory tracking control of a TLFM using a SMC. The point to be noted is that most of the reported SMC control techniques are applied to the linear dynamics of a TLFM. It is also seen from Table 4.1 that most SMC techniques have some disadvantages. The following paragraph explains the advantages and disadvantages of the reported SMC used to control a TLFM.

The normal SMC concept has some inherent disadvantages such as asymptotic stability in comparison with exponential stability, chattering and high gain [9], [10]. In the past decade, several efforts have been undertaken to improve the performance of a normal SMC [11–13]. The asymptotic stability problem is overcome with terminal sliding mode concept (TSMC) as noted in [14]. This method (TSMC) has some other advantages like high-precision performance and easy to use as reported in [7]. However, if the initial state of the system is far away from the equilibrium point, TSMC does not achieve a good convergence performance [15]. Reference [16] describes the singular perturbation problem using fuzzy nonsingular terminal sliding mode control for the slow subsystem and linear quadratic regulator (LQR) for the fast subsystem. The designed controller for the subsystems gives faster convergence and better tracking precision. In [17], a hybrid control scheme consisting of a continuous nonsingular terminal sliding mode controller and an observer based LQR controller has been proposed for the nonminimum phase problem of a flexible manipulator.

Another drawback of TSMC is the singularity problem, which necessitates a considerable control effort and can lead to chattering issues [15].

The major disadvantage with the normal SMC is the chattering issue. Many SMC techniques have been proposed to mitigate chattering, such as adaptive SMC [27], quasi-SMC, low pass filter method [28], fuzzy SMC [29], Lyapunov function approach [30] and higher order SMC (HOSMC) [31]. However, some techniques degrade the robustness and performances [15]. In [32], a second-order sliding curve and the new control law are designed for eliminating chattering present in the sliding surface for a rigid manipulator. Reference [33] discusses the first and the second-order reference models for variable structure system (VSS) control of compliant manipulators. The proposed VSS structure leads to smooth control actions that eliminate the chattering present in the normal variable structure control (VSC). Considering all aspects, the HOSMC is often considered superior in addressing chattering and improving control performance. In higher-order sliding mode control

Table 4.1: Categorisation of the reported SMC techniques used for control of two-link flexible manipulators (TLFMs).

Sl. No.	References of Paper	Types of SMC techniques used		Control problem	
1.	[18]	Terminal SMC + GA		Tip position control	
2.	[19]	Normal SMC +NN		Tip trajectory tracking control	
3.	[20]	Terminal SMC (TSMC)		Tip position control	
4.	[21]	Global TSMC		Trajectory tracking control	
5.	[22]	Non-singular TSMC + GA		Tip position control	
6.	[23]	Higher order non-singular TSMC + GA		Tip position control	
7.	[24]	Fuzzy non-singular TSMC +GA		Tip position control	
8.	[25]	Continuous non-singular TSMC + GA		Position control	
9.	[26]	Continuous non-singular TSMC combining HOSMC +GA		Position control	
		Using singular perturbation modelling method			
	References of Paper	Control technique	Control problem	Control technique	Control problem
13.	[16]	Fuzzy non-singular TSMC	Trajectory tracking	Reduced order-observer based LQR	State estimation and tip deflection suppression
14.	[17]	Continuous Non-singular TMC	Trajectory tracking	Reduced order-observer based LQR	State estimation and tip deflection suppression

(HOSMC), the chattering is effectively eliminated because the control signal is the continuous integration of its derivative.

Motivation: From the above discussion, it is evident that various combinations of sliding mode controllers with other control strategies have been proposed for a TLFM. However, achieving chattering-free robust stability for a nonlinear TLFM remains a relatively unexplored area. The design of a second order sliding mode controller specifically tailored for a TLFM serves as the motivation for this chapter.

4.2 Chapter objectives

- To design a sliding mode control using a conventional sliding surface.
- To design a second order sliding mode controller by using the normal sliding mode control designed above.
- To add the payload and the disturbances to the designed second order sliding mode controller and simulate the results.
- To compare the performances of the designed normal sliding mode controller and the proposed second order sliding mode controller with the payload and disturbances.

4.3 Structure of a sliding mode controller

The basic structure and design of a sliding mode controller is given in this section. Consider a single dimensional motion of unit mass as under:

$$\begin{cases} \dot{x}_1 = x_2 \\ \dot{x}_2 = u + f(x_1, x_2, t) \end{cases} \tag{4.1}$$

where u is the control input and $f(x_1, x_2, t)$ is the disturbance term. It is assumed that the disturbance is bounded with, $f(x_1, x_2, t) \leq L > 0$.

Here, the objective is to design a suitable controller so that both x_1 and x_2 converge to zero asymptotically in the presence of a bounded disturbance.

Let us introduce a new variable in the state space of the system in (4.1).

$$\sigma(x_1, x_2) = x_2 + cx_1, c > 0 \tag{4.2}$$

In order to achieve the asymptotic convergence of the state variables (x_1, x_2) to zero in the presence of the bounded disturbances $f(x_1, x_2, t)$, we need to drive the new variable σ to zero by means of control input u. This is achieved by applying Lyapunov theory to the σ-dynamics given in (4.3).

$$\dot{\sigma}(x_1, x_2) = cx_2 + f(x_1, x_2, t) + u, \sigma(0) = \sigma_0 \tag{4.3}$$

Consider a candidate of the Lyapunov function as follows:

$$V = \frac{1}{2}\sigma^2 \tag{4.4}$$

In order to achieve the asymptotic stability of (4.3), about the equilibrium point $\sigma = 0$, the following conditions must be satisfied:

- (a) $\dot{V} < 0$ for $\sigma \neq 0$,
- (b) $\lim_{|\sigma| \to \infty} V = \infty$

The derivative of V is computed as,

$$\dot{V} = \sigma\dot{\sigma} = \sigma(cx_2 + f(x_1, x_2, t) + u) \tag{4.5}$$

Assuming control input $u = -cx_2 + v$ and substituting it in (4.5) we can obtain it as,

$$\dot{V} = \sigma(f(x_1, x_2, t) + v) = \sigma f(x_1, x_2, t) + \sigma v \leq |\sigma|L + \sigma v \tag{4.6}$$

Considering $v = -\rho sign(\sigma)$, where $sign()$ is the signum function, with $\rho > 0$, using it in (4.6), we obtain

$$\dot{V} \leq |\sigma|L - |\sigma|\rho = -|\sigma|(\rho - L) = -\frac{\alpha}{\sqrt{2}}|\sigma| \tag{4.7}$$

where $\alpha = \frac{(\rho - L)}{\sqrt{2}}$ or control gain

$$\rho = L + \frac{\alpha}{\sqrt{2}} \tag{4.8}$$

Then, the control law u that drives ρ to zero is,

$$u = -cx_2 - \rho sign(\sigma) \tag{4.9}$$

The range of the derivative of the Lyapunov function $\dot{V}$ is,

$$\dot{V} \leq -\frac{\alpha}{\sqrt{2}}|\sigma| \leq -\alpha V^{1/2}, \alpha > 0 \tag{4.10}$$

Now, in order to achieve the finite time convergence (global finite time stability), condition (a) can be modified to the expression given in (4.10). The expression of the finite time can be obtained by separating the variables and integrating the inequality (4.10) over the finite time interval $0 \leq \tau \leq t$, to obtain,

$$\sqrt{V(t)} \leq -\frac{1}{2}\alpha t + \sqrt{V(0)} \tag{4.11}$$

Here, $V(t)$ reaches zero in a finite time t_r given in (4.11).

$$t_r \leq \frac{2\sqrt{V(0)}}{\alpha} \tag{4.12}$$

Remark 4.1. The dynamics of $\dot{\sigma}$ must be a function of control input u in order to successfully obtain the controller given in (4.8)–(4.9). This observation must be taken into account while designing the variable given in (4.2).

Remark 4.2. The first term L of the control gain (4.8) is designed to compensate the bounded disturbance $f(x_1, x_2, t)$, and the second term $\frac{\alpha}{\sqrt{2}}$ is responsible for determining the sliding surface reaching time given in (4.12). Larger the value of α, shorter the reaching time.

Definition 4.1: Equation (4.2) is a straight line in the state space of (4.1) and is called a sliding variable or surface.

Definition 4.2: The condition (4.10) is equivalent to

$$\sigma\dot{\sigma} \leq -\frac{\alpha}{\sqrt{2}}|\sigma| \tag{4.13}$$

and is called the reachability condition, which means that the trajectories of the system (4.1) are driven toward the sliding surface (4.2) and remain on it thereafter.

Definition 4.3: The control input $u(x_1, x_2)$ given in (4.9) that drives the states to the sliding surface (4.2) in finite time and keeps them on it thereafter in the presence of the bounded disturbance $f(x_1, x_2, t)$, is called a sliding mode controller.

For details regarding the design of the other variant of SMC and more theory, readers can refer to [34].

The next subsection discusses the design of a conventional sliding mode control for a two-link flexible manipulator.

4.4 Design of a conventional sliding mode control of a two-link flexible manipulator

This section describes the use of conventional sliding mode control for tip trajectory tracking control of TLFM dynamics (6.19). Let y_{di} be the desired trajectory to be tracked. Thus, the tip trajectory tracking errors are defined in (4.14), respectively.

$$e_i = y_i - y_{di} \tag{4.14}$$

where $i = 1, 2$

The second derivative of (4.14) is:

$$\ddot{e}_1 = \ddot{q}(1) + \frac{1}{l_1}(\phi_{11}\ddot{q}(3) + \phi_{12}\ddot{q}(4)) - \ddot{y}_{d1} \tag{4.15}$$

$$\ddot{e}_2 = \ddot{q}(2) + \frac{1}{l_2}(\phi_{21}\ddot{q}(5) + \phi_{22}\ddot{q}(6)) - \ddot{y}_{d2} \tag{4.16}$$

where $\frac{d}{dt}\ddot{q}(i)$, $(i = 1, ..., 6)$ are the elements of the column vector given in (6.25).

In order to guarantee the tracking of the desired trajectory and to bring the manipulator dynamics on the sliding surface, sliding surfaces defined in terms of errors are,

$$s_i = \dot{e}_i + c_i e_i \tag{4.17}$$

where c_i are the user defined positive constants.

Theorem 4.1. *Considering the conventional sliding surfaces defined in (4.17) and suppose that the control torque for the TLFM dynamics is defined as*

$$\begin{cases} \tau_2 = \frac{L_3L_4 - L_1L_6}{L_2L_4 - L_1L_5} - \rho_2 tanh(s_2) + \frac{L_1L_4}{L_1L_5 - L_2L_4}\rho_1 \tanh(s_1) \\ \tau_1 = \frac{L_3}{L_1} - (\frac{L_2}{L_1}\tau_2) - \rho_1 tanh(s_1) \end{cases} \tag{4.18}$$

where

$$\begin{cases} A_1 = \ddot{y}_{d1} - c_1\dot{e}_1 \\ A_2 = \ddot{y}_{d2} - c_2\dot{e}_2 \\ L_1 = N_{11} + \frac{\phi_{11}}{l_1}N_{31} + \frac{\phi_{11}}{l_1}N_{41} \\ L_2 = N_{12} + \frac{\phi_{11}}{l_1}N_{32} + \frac{\phi_{12}}{l_1}N_{42} \\ L_3 = A_1 + \ddot{q}A(1) + \frac{\phi_{11}}{l_1}\ddot{q}A(3) + \frac{\phi_{12}}{l_1}\ddot{q}A(4) \\ L_4 = N_{21} + \frac{\phi_{21}}{l_2}N_{51} + \frac{\phi_{22}}{l_2}N_{61} \\ L_5 = N_{22} + \frac{\phi_{21}}{l_2}N_{52} + \frac{\phi_{22}}{l_2}N_{62} \\ L_6 = A_2 + \ddot{q}A(2) + \frac{\phi_{21}}{l_2}\ddot{q}A(5) + \frac{\phi_{22}}{l_2}\ddot{q}A(6) \end{cases} \tag{4.19}$$

where $N(i, j)$ are the elements of $N = M^{-1}(q)$ and $i = 1, ..., 6$. $\ddot{q}A = N(h + Kq + D\dot{q})$, $\ddot{q}A(i)$ are the elements of $\ddot{q}A$. Then, the trajectories of the flexible manipulator converge to the sliding surface (4.17) in finite time and remain on it.

Proof. Consider a Lyapunov function candidate as

$$\begin{cases} V_1 = \frac{1}{2}s_1^2 \\ \dot{V}_1 = s_1\dot{s}_1 \\ \dot{V}_1 = s_1[\ddot{e}_1 + c_1\dot{e}_1] \end{cases} \tag{4.20}$$

From the first equation of (4.15), (4.20) is written as

$$\dot{V}_1 = s_1[\ddot{q}(1) + \frac{1}{l_1}(\phi_{11}\ddot{q}_3 + \phi_{12}\ddot{q}(4)) - \ddot{y}_{d1} + c_1\dot{e}_1] \tag{4.21}$$

Now, using (4.19), (4.21) is written as

$$\dot{V}_1 = s_1[L_1\tau_1 + L_2\tau_2 - L_3] \tag{4.22}$$

Using (4.18) in (4.22),

$$\dot{V}_1 = s_1[-L_1\rho_1 tanh(s_1)] \leq -L_1\rho_1|s_1| \tag{4.23}$$

Similarly for the second sliding surface (s_2), we can show the stability by considering another Lyapunov function of the form, $V_2 = \frac{1}{2}s_2^2$. The time derivative of V_2 using (4.16), (4.18), (4.19) is

$$\dot{V}_2 \leq -(\frac{L_1L_5 - L_2L_4}{L_1})\rho_2|s_2| \tag{4.24}$$

The terms L_1 and $\frac{L_1L_5-L_2L_4}{L_1}$ are the time varying positive quantities and ρ_1, ρ_2 are the positive constants. Hence, using the Lyapunov stability theory we can show that $\dot{V}_1(s_1)$ and $\dot{V}_2(s_2)$ are negative definite. Therefore, the existence of the manipulator dynamics (3.47) on the sliding surface is ensured. □

4.5 Design of a second order sliding mode control of a two-link flexible manipulator

Suppose y_{di} is the twice differentiable desired tip trajectory for the TLFM (6.19). The tip trajectory tracking error using (6.19) can be defined as

$$e_i = y_i - y_{di} \tag{4.25}$$

The third derivative of (6.22) can be written as:

$$\frac{d}{dt}(\ddot{e}_1) = -\frac{d}{dt}(\ddot{y}_{d1}) + \{\frac{d}{dt}\ddot{q}(1) + \frac{1}{l_1}(\phi_{11}(\frac{d}{dt}\ddot{q}(3)) + \phi_{12}(\frac{d}{dt}\ddot{q}(4)))\} \tag{4.26}$$

$$\frac{d}{dt}(\ddot{e}_2) = -\frac{d}{dt}(\ddot{y}_{d2}) + \{\frac{d}{dt}\ddot{q}(2) + \frac{1}{l_2}(\phi_{21}(\frac{d}{dt}\ddot{q}(5)) + \phi_{22}(\frac{d}{dt}\ddot{q}(6)))\} \tag{4.27}$$

where $\frac{d}{dt}\ddot{q}(i)$, $(i = 1, ..., 6)$ are the elements of the column vector given in (6.25).

$$\frac{d}{dt}(\ddot{q}) = \{\dot{B}^{-1}(q)[-h(q) - Kq - D\dot{q} + b\tau]\} + \{B^{-1}(q)[-\dot{h}(q) - K\dot{q} - D\ddot{q} + b\dot{\tau}]\} \tag{4.28}$$

Now, the aim is to design a suitable second-order SMC to operate on the manipulator dynamics (6.19) such that the TLFM follows the desired trajectories along with suppression of tip deflection.

Now, a second-order sliding mode control (SO-SMC) is designed for the tip trajectory tracking control of the TLFM. The SO-SMC is designed in terms of the conventional first order SMC (4.17). The second-order sliding surface is designed as

$$\sigma_i = s_i + \lambda_i \dot{s}_i \quad for\ i = 1, 2 \tag{4.29}$$

where λ_i is a positive constant, whose value is user dependent. The necessary condition for the existence of a second-order sliding surface is to satisfy $\dot{\sigma} = 0$. Thus, the equivalent sliding mode dynamics of (4.29) can be defined as

$$\ddot{s}_i = -K_{Mi}\dot{s}_i \tag{4.30}$$

where $K_{Mi} = 1/\lambda_i$

Theorem 4.2. *The equivalent sliding mode dynamics obtained in (4.30) is asymptotically stable and manipulator trajectories converge to the sliding surface $\sigma = 0$.*

Proof. Consider a Lyapunov function candidate as $V(s_i) = \frac{1}{2}\dot{s}_i\dot{s}_i^T$, whose derivative can be written using (4.30) as

$$\dot{V}(s_i) = -K_M\dot{s}_i^2 \tag{4.31}$$

$\dot{V}(s_i)$ in (4.31) is a negative definite function. Thus, the sliding surface defined in (4.29) is stable and satisfies $\sigma = \dot{\sigma} = 0$, and hence, manipulator dynamics follow the desired trajectories. The existence of the manipulator dynamics (6.19) on the sliding surface (4.29) is achieved using Theorem 4.3. □

Theorem 4.3. *Considering the second-order sliding surface defined in (4.29) and supposing that the control torque for the TLFM dynamics (6.19) is defined as*

$$\dot{\tau}_1 = \frac{Z_2}{Z_3} - \frac{Z_4}{Z_3}\dot{\tau}_2 - \rho_1 tanh(\sigma_1) \tag{4.32}$$

$$\dot{\tau}_2 = \frac{Z_2Z_7 - Z_3Z_6}{Z_4Z_7 - Z_3Z_8} - Z_3\rho_2 tanh(\sigma_2) + \frac{Z_7Z_3}{Z_3Z_8 - Z_4Z_7}\rho_1 tanh(\sigma_1) \tag{4.33}$$

such that

$$\begin{cases} N = M^{-1}(q) \\ A1 = \dot{N}(\tau - h - Kq - D\dot{q}) \\ A2 = N(\dot{h} + K\dot{q} + D\ddot{q}) \\ A3 = A1 - A2 \\ Z_1 = -K_{M1}\dot{s}_1 + \frac{d}{dt}\ddot{y}_{d1} - c_1\ddot{e}_1 \\ Z_2 = Z_1 - A_3(1) - \phi'_{11}A_3(3) - \phi'_{12}A_3(4) \\ Z_3 = N_{11} + \phi'_{11}N_{31} + \phi'_{12}N_{41} \\ Z_4 = N_{12} + \phi'_{11}N_{32} + \phi'_{12}N_{42} \\ Z_5 = -K_{M2}\dot{s}_2 + \frac{d}{dt}\ddot{y}_{d2} - c_2\ddot{e}_2 \\ Z_6 = Z_5 - A_3(2) - \phi'_{21}A_3(5) - \phi'_{22}A_3(6) \\ Z_7 = N_{21} + \phi'_{21}N_{51} + \phi'_{22}N_{61} \\ Z_8 = N_{22} + \phi'_{21}N_{52} + \phi'_{22}N_{62} \end{cases} \tag{4.34}$$

where $N(i,j)$ are the elements of $N = M^{-1}(q)$, $A_3(i)$ are the elements of A_3 and $i = 1, ..., 6$. ρ_i is the user dependent positive constant gain. Then, the trajectories of the flexible manipulator converge to the sliding surface (4.29) in finite time and remain on it. The actual value of the torque input can be found by the integration of $\dot{\tau}_i$.

Proof. Consider another Lyapunov function candidate as

$$V_2(\sigma_i) = \frac{1}{2}(\sigma_1^2 + \sigma_2^2) \tag{4.35}$$

Taking the derivative of (4.35) and using (4.32), (4.33), the time derivative of (4.35) can be written as

$$\dot{V}_2(\sigma_i) = -Z_3\rho_1|\sigma_1| - \rho_2|\sigma_2| \tag{4.36}$$

where Z_3 is a time varying positive quantity and ρ_1, ρ_2 are the user dependent positive quantities. Thus, according to the Lyapunov stability theory $\dot{V}_2(\sigma_i)$ is negative definite. Hence, joint trajectories of the flexible manipulator (6.19) follow the desired trajectory using the control input torque proposed in (4.32), (4.33). □

4.6 Results and discussion

The parameter values of a physical TLFM used for simulating the flexible manipulator dynamics (6.19) is given in Table 3.1. In the following, we pursue the payload of the 0.145 kg condition. The expression of the desired trajectories used for link-1 and link-2 are given as.

$$\begin{cases} y_{d1} = \frac{\pi}{4} - \frac{7}{6}e^{-\frac{3}{2}t} + \frac{7}{19}e^{-2t} \\ y_{d2} = \frac{\pi}{6} - \frac{7}{6}e^{-\frac{3}{2}t} + \frac{7}{11}e^{-2t} \end{cases} \tag{4.37}$$

In this chapter, all the simulations are carried out using $ode - 45$ solver in a MATLAB simulation environment. The initial conditions for simulating the TFLM dynamics (6.19), control inputs (4.32), (4.33) are considered as:

$$\begin{cases} q(0) = [0.1, 0.1, 0, 0.0001, 0, 0.0001]^T \\ \dot{q}(0) = [0, 0, 0, 0, 0, 0]^T \\ \tau_i(0) = [0, 0]^T \end{cases} \tag{4.38}$$

Some other constants used for simulation of sliding surfaces (4.17) and (4.29) are $\rho_1 = \rho_2 = 7,\ c_1 = 7,\ c_2 = 7,\ \lambda_1 = \lambda_2 = 5$.

Remark 4.4. The constants in (4.17) are chosen in a manner to achieve a good tracking performance and small control effort. Initially, we have chosen the value of c_1 and c_2 arbitarily and checked the IAE of tip trajectory tracking as well as $\|u\|_2$. We have changed the value of the parameters c_1 and c_2 by trial and error till we found a set of suitable values which give a lower tip trajectory tracking error along with lower $\|u\|_2$.

4.6.1 Results of SO-SMC

The tip trajectory tracking responses for both the links are shown in Fig. 4.1. The modes of the first and second flexible links with 0.145 kg payload are shown in Fig. 4.2. Figure 4.2 depicts the first two modes of link-1 and link-2. The second mode of both the links has lesser magnitude of deflection than the first mode of both the links. The responses of the tip deflection of the links are shown in Fig. 4.3. First order sliding surfaces designed here are shown in Fig. 4.4. The responses of the proposed second-order sliding surfaces are shown in Fig. 4.5. It is seen from Fig. 4.1 that the tip trajectory tracking using SO-SMC is achieved within 0.6 s. It is apparent from Fig. 4.3 that the tip deflection of the first link is suppressed within 0.02 mm, and tip deflection of the second link

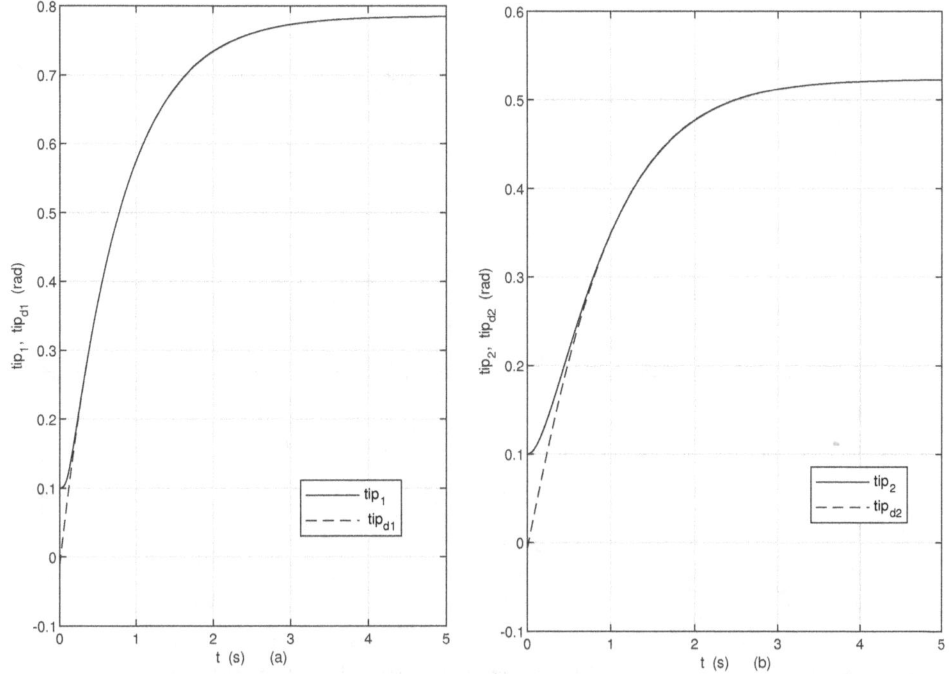

Figure 4.1: Responses of the tip trajectory tracking for: (a) link-1 and (b) link-2.

is suppressed within 0.002 mm. It is seen from Fig. 4.5 that the reaching time for the second-order sliding surface for link-1 and link-2 is within 2.8 s and 3.8 s. It is clear from Figs. 4.4 and 4.5 that the chattering is removed and hence tip trajectory tracking control and tip deflection suppression is achieved effectively.

4.6.2 Comparison of SO-SMC and SMC

In this section, we compare the performance of SO-SMC with that of SMC. The parametric constants are chosen in such a manner that both the controllers have the same sliding surface constants, i.e., $c_1 = c_2 = 7$. Tip trajectory tracking errors using both the controllers (SO-SMC and SMC) are shown in Fig. 4.6. It is visible from Fig. 4.6 that the tracking time for the link-1 using SO-SMC is comparatively less as compared with the SMC. In case of link-2 tracking time is almost the same. Tip deflections using both the controllers (SO-SMC and SMC) are shown in Fig. 4.7. It is apparent from Fig. 4.7 that the tip deflection for both the links using the SO-SMC is less as compared with normal SMC. Responses to the required torque inputs using SO-SMC and SMC controllers are shown in Fig. 4.8. It is noted from Fig. 4.8 that the required torque in case of SO-SMC is comparatively less compared with the SMC. It is also visible from Fig. 4.8 that the torque input using SO-SMC is smoother as compared with the normal SMC. The comparison between the performance indices of SMC and SO-SMC are shown in Table 4.2. The performances of the control techniques are evaluated using the following performance indices.

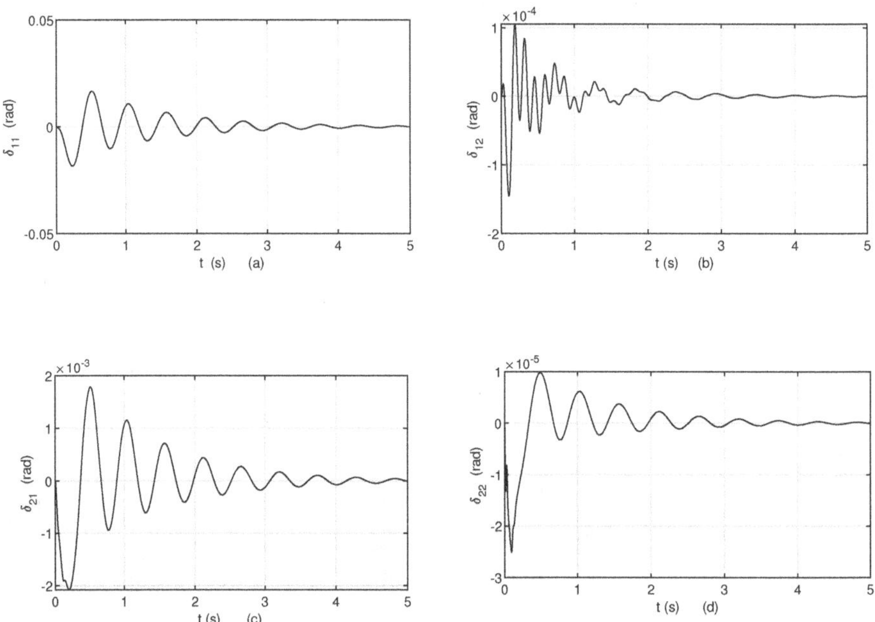

Figure 4.2: Modes of the links of the TLFM during tip trajectory tracking of: (a) the first mode of link-1, (b) the second mode of link-1, (c) the first mode of link-2 and (d) the second mode of link-2.

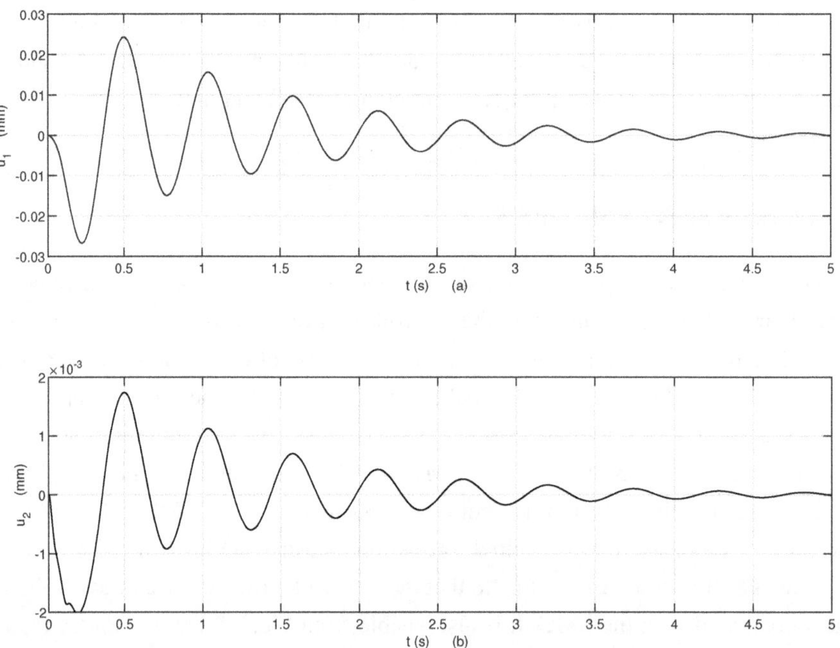

Figure 4.3: Responses of the tip deflections during tip trajectory tracking of: (a) link-1 and (b) link-2.

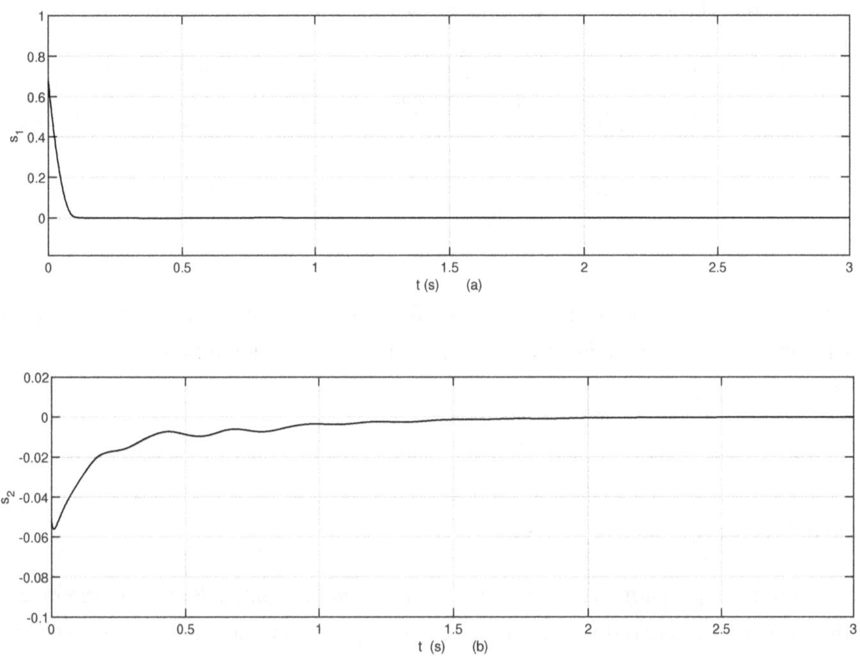

Figure 4.4: First order sliding surfaces used for the tip trajectory tracking control of: (a) link-1 and (b) link-2.

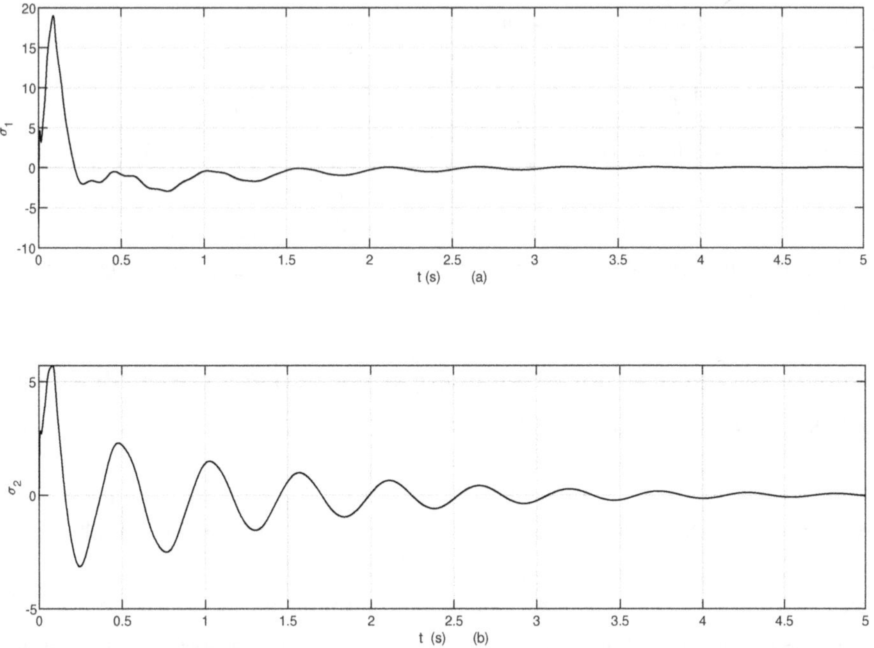

Figure 4.5: Second-order sliding surfaces used for the tip tracking control of: (a) link-1 and (b) link-2.

Table 4.2: Comparison of the performance indices between the controllers.

Control technique	link	IAE of tip trajectory tracking	$\|\|u_i\|\|_2$ of tip deflection	Total Variance (TV)
SMC	link-1	0.01936	0.682	31.0281
	link-2	0.0125	0.032	13.1874
SO-SMC	link-1	**0.0081**	**0.581**	**9.6910**
	link-2	**0.0102**	**0.028**	**7.5833**

1. Tracking performance/output performance: To evaluate the tracking responses of the controllers, an integral absolute error (IAE) index is used. It is shown in (4.39).

$$IAE = \int_{0}^{T_f} |e(t)|dt \tag{4.39}$$

where $e(t)$ is the tip trajectory tracking error.

2. Tip deflection suppression performance: The tip deflection suppression performance of both the links is calculated in terms of the required energy consumption and is expressed as $||u_i||_2$, where i is the serial number of the link.

3. Control performances: The control input performances of the controllers are evaluated using a total variation (TV) parameter and 2-norm of control input(n_2). The expression of a TV parameter and 2-norm of the control input are given as,

$$\begin{cases} TV = \sum_{i=1}^{r} ||\tau_{j(i+1)} - \tau_{ji}||, \; j = no.\, of joints \\ n_2 = ||\tau_j||_2 \end{cases} \tag{4.40}$$

Here, a TV is a measure of the smoothness of a signal and is expected to have a low value [35]. 2-norm of the control input expresses the required energy and is expected to be as low as possible [35].

4.6.3 Comparison of SO-SMC with nominal payload and SO-SMC with 0.3 kg payload along with external disturbance

In this section, in order to check the robustness of the proposed controller, a comparison is done between the proposed SO-SMC and the SO-SMC with additional payload which adds up to 0.3 kg along with a matched disturbance, where $d_1 = 0.5\sin(\theta_1)\sin(\dot{\theta}_1)$, $d_2 = 0.5\sin(\theta_2)\sin(\dot{\theta}_2)$. The tip trajectory tracking error, tip deflections and the control inputs are shown in Figs. 4.9, 4.10 and 4.11, respectively. In Fig. 4.9, it is seen that when an additional payload is added along with the disturbances, the tip trajectory tracking errors remain the same in both the cases of the proposed controllers. But, in the cases of Fig. 4.10 and 4.11, the SO-SMC with the additional payload and disturbances, the tip deflections as well as the torque have a negligible increment. Thus, the designed SO-SMC is robust against the addition of payload and disturbances.

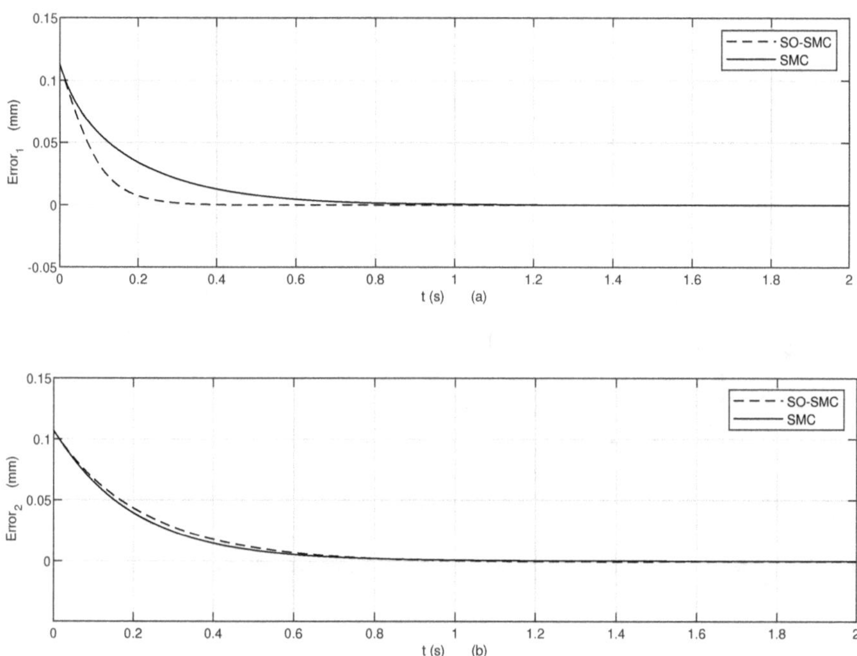

Figure 4.6: Tip trajectory tracking errors using the SO-SMC and conventional SMC.

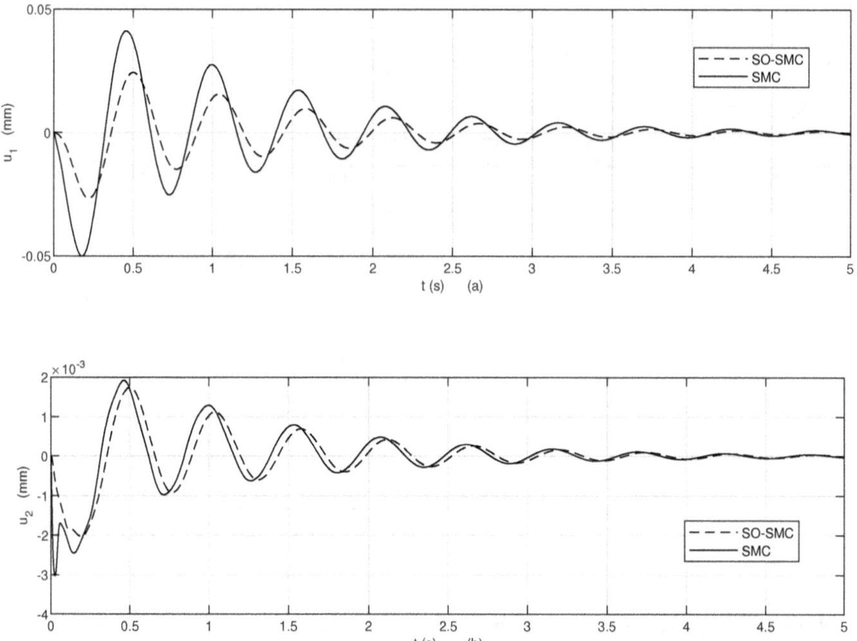

Figure 4.7: Tip deflections using the SO-SMC and conventional SMC for: a) link-1 and b) link-2.

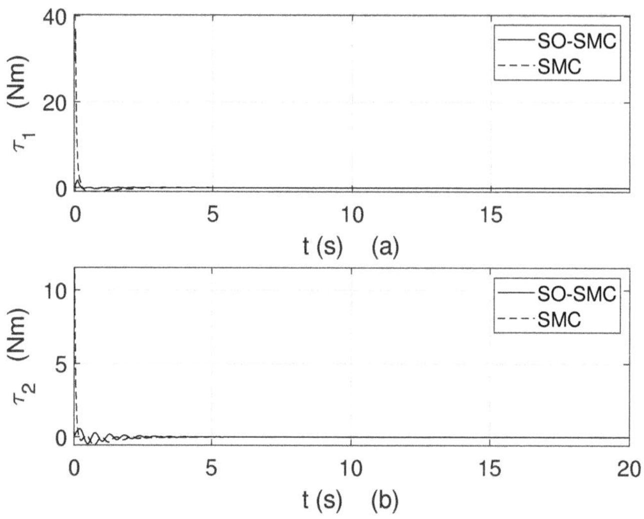

Figure 4.8: Required control efforts using the SO-SMC and conventional SMC for: a) link-1 and b) link-2.

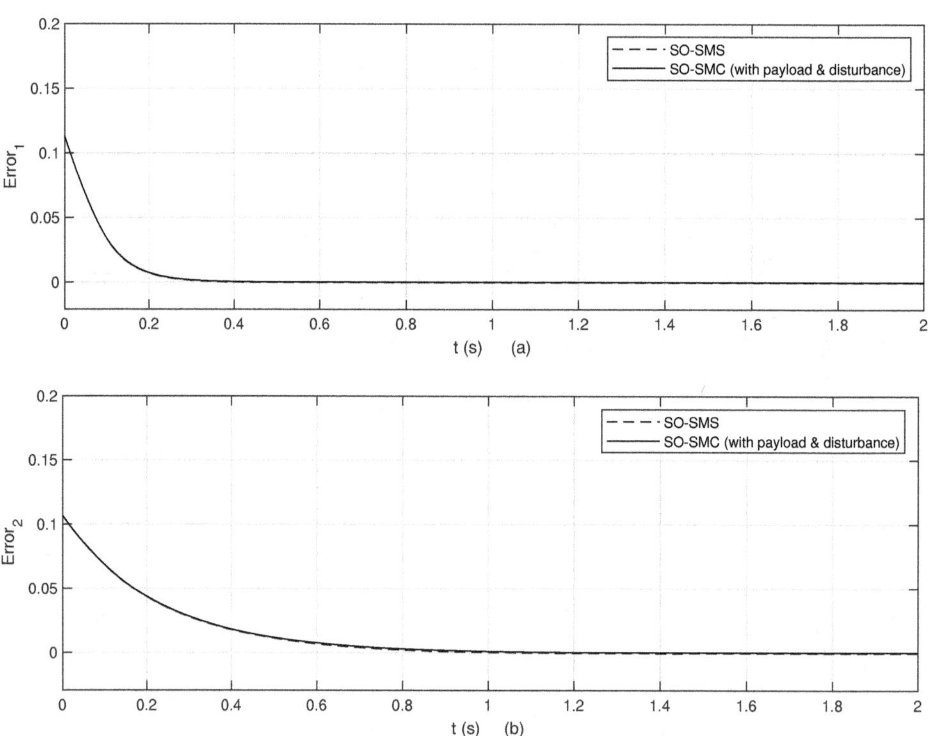

Figure 4.9: Tip trajectory tracking errors using the SO-SMC and SO-SMC with payload and disturbances for: a) link-1 and b) link-2.

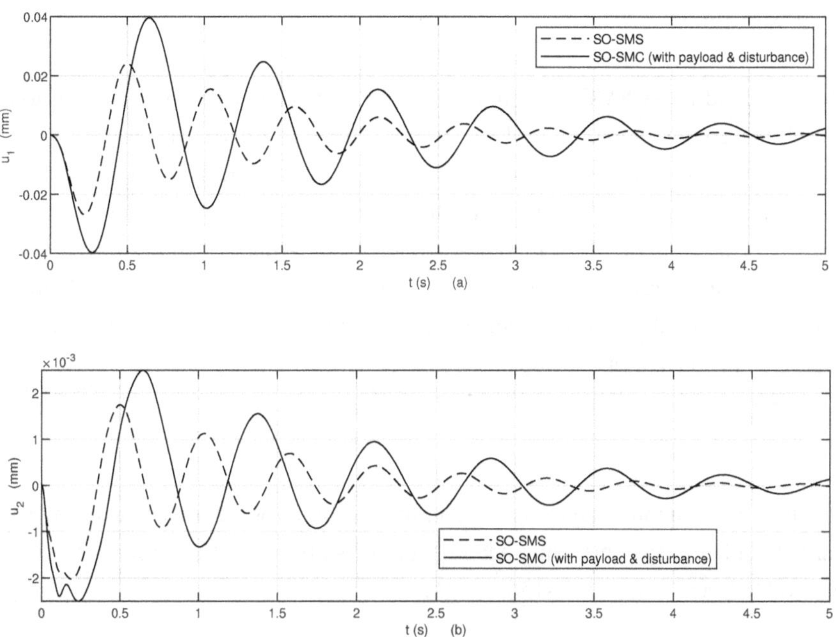

Figure 4.10: Tip deflection using the SO-SMC and SO-SMC with payload and disturbance for: a) link-1 and b) link-2.

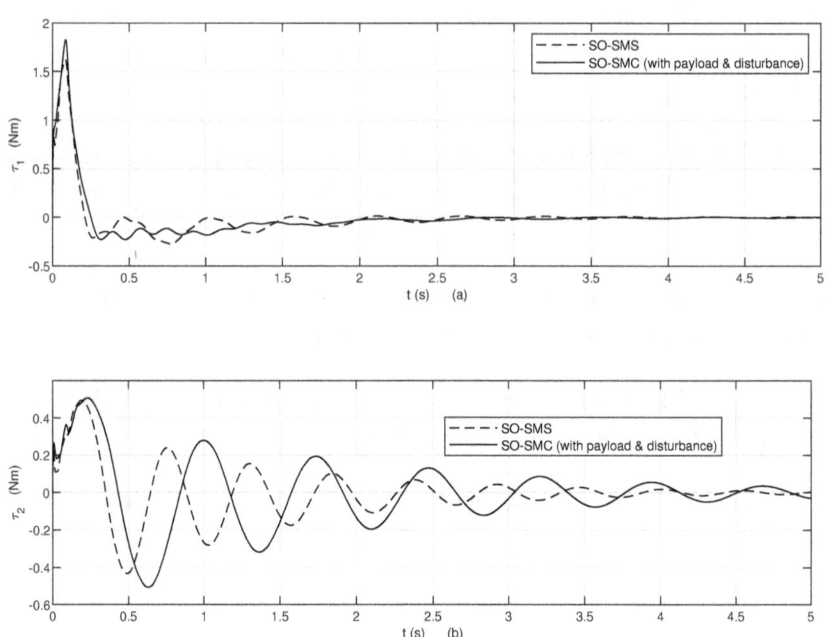

Figure 4.11: Required control efforts using the SO-SMC and SO-SMC with payload and disturbance for: a) link-1 and b) link-2.

4.7 Chapter summary

In this chapter, a second-order SMC is designed for the tip trajectory tracking control of a planar nonlinear TLFM along with suppression of the tip deflection. The nonlinear model of the TLFM is derived using the assumed modes method (AMM) with the consideration of two modes. First, a conventional SMC is designed in terms of tip tracking error, and then a second order SMC is designed in terms of conventional SMC. It is shown that the chattering in sliding surfaces and control inputs are removed. Fast tip trajectory tracking, quick tip deflection suppression and chattering free control inputs are considered as the performance metrics of the designed SO-SMC and is found better in comparison to conventional SMC.

References

[1] S. B. Pradhan, S. K. Stability analysis and controller design for the performance improvement of disturbed nonlinear systems using adaptive global sliding mode control approach. *Nonlinear Dyn.*, vol. 83, no. 3, pp. 1557–1565, 2016.

[2] C. T. Kiang, A. Spowage, and C. K. Yoong. Review of Control and Sensor System of Flexible Manipulator. *Review of Control and Sensor System of Flexible Manipulator*, vol. 77, no. 1, pp. 187–213, 2014.

[3] C. T. Kiang, A. Spowage, and C. K. Yoong. Review of Control and Sensor System of Flexible Manipulator. *Review of Control and Sensor System of Flexible Manipulator*, vol. 77, no. 1, pp. 187–213, 2014.

[4] S. B. Pradhan, S. K. Nonlinear Adaptive Model Predictive Controller for a Flexible Manipulator: An Experimental Study. *IEEE Trans. Control Syst. Technol.*, vol. 22, no. 5, pp. 1754–1768, 2014.

[5] Y. Yu, Y. Yuan, X. Fan, and H. Yang. Back-stepping control of two-link flexible manipulator based on extended state observer. *Adv. Sp. Res.*, vol. 56, pp. 2312–2322, 2015.

[6] B. Subudhi and A. S. Morris. Soft computing methods applied to the control of a flexible robot manipulator. *IEEE Trans. Appl. Soft Comput.*, 2009.

[7] M. Galicki. Finite-time trajectory tracking control in a task space of robotic manipulators. *Automatica*, vol. 67, pp. 165–170, 2016.

[8] H. K. Khalil. *Nonlinear systems*. Pearson New International, 3rd edition ed., 2002.

[9] S. Kamal, J. A. Moreno, A. Chalanga, and F. L. M. Bandyopadhyay, B. Continuous terminal sliding-mode controller. *Automatica*, 2016.

[10] C. S. Chiu. Derivative and integral terminal sliding mode control for a class of MIMO nonlinear systems. *Automatica*, 2012.

[11] S. Ding, A. Levant, and S. Li. Simple homogeneous sliding-mode controller. *Automatica*, 2016.

[12] A. Pisano, M. Tanelli, and A. Ferrara. Automatica Switched / time-based adaptation for second-order sliding mode. *Automatica*, vol. 48, no. 2, pp. 308–326, 2012.

[13] Q. Zhu, X. Yu, A. Song, S. Fei, Z. Cao, and Y. Yang. Automatica On sliding mode control of single input Markovian jump systems. *Automatica*, vol. 68, pp. 286–293, 2014.

[14] Y. Feng, X. Yu, and F. Han. On nonsingular terminal sliding-mode control of nonlinear systems. *Automatica*, vol. 49, no. 6, pp. 1715–1722, 2013.

[15] X. T. Tran and H. J. Kang. Adaptive hybrid High-Order terminal sliding mode control of MIMO uncertain nonlinear systems and its application to robot manipulators. *Int. J. Precis. Eng. Manuf.*, vol. 16, no. 2, pp. 255–266, 2015.

[16] Y. Wang, Y. Feng, and X. Yu. Fuzzy terminal sliding mode control of two-link flexible manipulators. In *In: 34th IEEE Annual Conference of Industrial Electronics*, pp. 1620–1625, 2015.

[17] Y. Wang, F. Han, Y. Feng, and H. Xia. Hybrid Continuous Nonsingular Terminal Sliding Mode Control of Uncertain Flexible Manipulators. In *In: 40th IEEE Annual Conference of the Industrial Electronics Society (IECON)*, pp. 190–196, 2014.

[18] N. Zhang, Y. Feng, and X. Yu. Optimization of Terminal Sliding Control For Two-Link Flexible Manipulators. In *In: The 30th Annual Conference of the IEEE Industrial Electronics Society*, pp. 1318–1322, 2004.

[19] Y. Tang, F. Sun, and Z. Sun. Neural network control of flexible-link manipulators using sliding mode. *Neurocomputing*, vol. 70, no. 1, pp. 288–295, 2006.

[20] W. Dongmei. The design of terminal sliding controller of two- link flexible manipulators. In *In: International Conference on Control and Automation*, pp. 733–737, 2007.

[21] C. Ming, J. Qing-Xuan, and S. Han-Xu. Global Terminal Sliding Mode Robust Control for Trajectory Tracking and Vibration Suppression of Two-Link Flexible Space Manipulator. In *In: IEEE International Conference on Intelligent Computing and Intelligent Systems*, pp. 353–357, 2009.

[22] Y. Feng and X. Yu. Nonsingular Terminal Sliding Mode Control of Uncertain Two-Link Flexible Manipulators. In *In: 35th Annual Conference of IEEE Industrial Electronics*, pp. 2295–2300, 2009.

[23] Y. Wang, C. Wang, P. Lu, and Y. Wang. High-order nonsingular terminal sliding mode optimal control of two-link flexible manipulators. In *In: 37th IEEE Annual Conference of the Industrial Electronics Society (IECON 2011)*, pp. 3953–3958, 2011.

[24] Y. Wang, H. Xia, and C. Wang. Hybrid Controllers for Two-Link Flexible Manipulators. In *In: ICAIC 2011, Part III*, pp. 409–418, 2011.

[25] Y. Wang and L. Sun. On the Optimized Continuous Nonsingular Terminal Sliding Mode Control of Flexible Manipulators. In *In: 4th International Conference on Instrumentation and Measurement, Computer, Communication and Control*, pp. 324–329, 2014.

[26] Y. Wang, C. Yuqing, and X. Hongwei. Optimized Continuous Nonsingular Terminal Sliding Mode Control of Uncertain Flexible Manipulators. In *In: Proceedings of the 34th Chinese Control Conference*, pp. 3392–3397, 2015.

[27] H. Li, P. Shi, and L. Yao, D. and Wu. Observer-Based Adaptive Sliding Mode Control for Nonlinear. *Automatica*, vol. 64, pp. 536–544, 2016.

[28] H. Lee and V. I. Utkin. Chattering Suppression in Sliding Mode Control System, journal = Annu. Rev. Control, year = 2007, volume = 31, pages = 179-188, number = 2.

[29] M. Roopaei and M. Z. Jahromi. Chattering-free fuzzy sliding mode control in MIMO uncertain systems. *Nonlinear Anal. Theory, Methods Appl.*, vol. 71, no. 10, pp. 4430–4437, 2009.

[30] A. Polyakov, D. Effimov, and W. Perruquetti. Finite-time and fixed-time stabilization: Implicit Lyapunov function approach. *Automatica*, vol. 51, no. 10, pp. 332–340, 2015.

[31] A. Pisano, M. Tanelli, and A. Ferrara. Automatica Switched / time-based adaptation for second-order sliding mode. *Automatica*, vol. 64, pp. 126–132, 2016.

[32] J. A. T. Machado and J. L. Martins de Carvalho. A new variable structure control for robot manipulators. In *In: ISIC'88-Third IEEE International Symposium on Intelligent Control*, pp. 441–446, 1988.

[33] J. A. T. Machado. Variable structure control of manipulators with compliant joints. In *In: ISIE'93-IEEE International Symposium on Industrial Electronics*, pp. 441–446, 1993.

[34] Y. Shtessel, C. Edwards, L. Fridman, and A. Levant. *Sliding mode control and observation*. Springer Science, 1st edition ed., 2014.

[35] T. Xuan-Tao and K. Hee-Jun. Adaptive Hybrid High-Order Terminal Sliding Mode Control of MIMO Uncertain Nonlinear Systems and its Application to Robot Manipulators. *International Journal of Precision Engineering and Manufacturing*, vol. 16, pp. 255–266, 2015.

Chapter 5

Design of Controllers for a TLFM using the Singular Perturbation Technique

In this chapter, the dynamics of the TLFM are divided into two parts using the two-time scale separation principle (singular perturbation), namely the slow subsystem involving the rigid parts and the fast subsystem which incorporates the flexible dynamics. Thus, control design of the TLFM becomes simple and effective by designing a LMI based sliding mode controller for the slow subsystem and a LMI based state feedback controller for the fast subsystem. These above two subsystem controllers are then superimposed to achieve a composite controller with the necessary boundary layer correction to take care of the two time scale separation of the complex dynamics of the TLFM. Similarly, a second composite controller is designed where an adaptive SMC is designed for the slow subsystem and a backstepping controller is designed for the fast subsystem.

The chapter is organised with an introduction, chapter objective and the model decomposition by the assumed modes method using singular perturbation. The design of the composite controllers for the slow and fast subsystems using the mathematical model derived in Chapter 3 is presented. The obtained simulation results are analysed, and finally, the conclusion of this chapter is provided.

5.1 Introduction

As discussed in Chapter 2, controlling a flexible manipulator is difficult due to its distributed link flexibility which makes the system noncollocated and underactuated. Further, controlling a flexible manipulator becomes more difficult while handling variable payloads and in the presence of disturbances. So, in order to handle variable payloads and to achieve a good tip trajectory tracking along with the suppression of the tip deflections, a good controller as well as an enhanced modelling method should be employed to provide appropriate control torques to actuators to achieve the said control problem.

The design of a controller largely depends on the modelling method used to model the system. Many modelling methods are available in the literature for the modelling of flexible manipulators. The commonly used modelling methods are assumed modes method [1], lumped parameters method and finite element method [2] as discussed in Chapter 2. As discussed, among these three modelling methods, assumed modes method (AMM) is widely used. The AMM gives the desirable response of the flexible dynamics by suitable choice of the number of modes. Another interesting modelling method is modelling of system by assumed modes method with the singular perturbation (SP) technique. In the singular perturbation technique a two-time scale separation principle is used [1, 3]. In the case of flexible manipulators, the singular perturbation is used to divide the manipulator dynamics into a slow subsystem consisting of the rigid dynamics and a fast subsystem consisting of the flexible dynamics [3]. Thus, it is easy to design two separate control inputs for each subsystem and the desired performances can be achieved.

The use of a singular perturbation (SP) modelling method for the design of a control technique for quick tip deflection suppression is more worthy as compared with other methods because in SP, the dynamics of the fast subsystem representing the flexible motion of FMs can be used to design a separate control technique for the quick suppression of the deflection of links.

Motivation 1:

Most of the reported control techniques on TLFMs use the assumed modes modelling method for designing their controllers. But, designing of a controller for a two-link flexible manipulator with singular perturbation using assumed modes method is found to be predominantly less. The reported papers using SP on the dynamics of a TLFM are classified in Table 5.1. It is seen from Table 5.1, that the use of singular perturbation for designing a controller of a TLFM is still less explored. Motivated with the above discussion, this chapter attempts to design some control techniques by the use of the singular perturbation technique through the assumed modes method for a TLFM.

Motivation 2:

It is seen from Table 5.1, that various types of control techniques are used like PID, PID with state feedback control, LQR and soft computing techniques. Among the aforesaid controllers, SMC gives better results due to its inherent advantages like robustness to uncertainities and insensitivity to disturbances. Generally, SMC ensures global stability and convergence of system dynamics [17], [18], [19].

It is also observed from Table 5.1 that various combination of SMC are used such as normal SMC, variable structure control, fuzzy non-singular TSMC, adaptive normal SMC with H_∞. But, the main challenge involved in SMC is the selection of the sliding surface gains. All the above reported papers

Table 5.1: Categorisation of the controllers applied on a singular perturbation model of two-link flexible manipulators.

References of paper	Types of control (Slow subsystem)	Types of control (Fast subsystem)
[4]	PID Feedback control	PID
[5]	Feedback linearisation	Linear quadratic regulator (LQR)
[1]	Computed torque control	LQR based state feedback
[6]	PD	Feedback control
[7]	VSC	Virtual force control
[8]	PID+ANN	H_∞ control
[9]	Adaptive Normal SMC with H_∞	LQR
[10]	PID Feedback control	PID
[11]	Fuzzy non-singular TSMC	Reduced order-observer based LQR
[12]	PD type inverse dynamic based control	Lyapunov based control
[13]	Normal SMC	H_∞
[14]	VSC	Lyapunov based controller
[15]	NN	LQR
[16]	Continuous non-singular TMC	Reduced order-observer based LQR
	LMI-SMC	LMI-state feedback controller
	Adaptive SMC	Backstepping

used for controlling TLFM using SMC, select the sliding surface gains depending on the choice of the user. It is also observed that the combination of LMI-SMC and LMI state feedback controller is not found in the literature. Therefore, finding and obtaining a reasonable optimised value for the sliding surface gains as well as the gains of the state feedback controller for tip trajectory tracking and tip deflection control, respectively, using LMI is one of the motivations of this chapter. Hence, a composite control is designed using LMI based SMC for the slow subsystem for tip trajectory tracking control and LMI based state feedback controller for the fast subsystem for tip deflection suppression. The stability analysis of the designed controller is shown using Lyapunov stability analysis.

Many control problems are considered in the literature for a flexible manipulator. The most commonly used control problems are path planning/trajectory tracking for the hub angle and path planning/trajectory tracking for the tip position. Various desired paths/trajectories are considered in the literature. The available desired signals used for the flexible manipulator in the literature are listed in Table 5.2.

Motivation 3:

1. It is seen from Table 5.2 that the use of a signal generated from a 3-D chaotic system as the desired trajectory for a flexible manipulator is not found in the literature. Hence, this

Table 5.2: Type of desired trajectories used for the trajectory tracking control of a flexible manipulator.

Sl. No.	Desired trajectories	References of papers
1.	Bang-bang	[20]
		[21]
2.	Circular	[22]
		[23]
3.	Exponentially varying	[7]
		[24]
4.	Straight Link	[8]
5.	Chaotic signal	

work uses a chaotic signal as the desired trajectory for the path planning/trajectory tracking control of a two-link flexible manipulator.

2. From Table 5.1 and from the second motivation it can be discussed that the control law gains are determined using LMI. However, selecting the appropriate gains of the sliding mode control law may be difficult when the TLFM is subjected to variable payloads, when the system is required to track the unpredictable (chaotic) desired trajectory and in the presence of external disturbances. Thus, there is a need to adapt the gain of the sliding mode control law.

3. Also, it is expected that the link deflection of the TLFM increases when the system is required to track the unpredictable (chaotic) desired trajectory and in the presence of external disturbances. Therefore, there is a need to design a robust controller to suppress the increasing tip deflections.

4. Motivated by the previous two points, we design a composite controller for tracking the chaotic desired trajectory in the presence of variable payloads and external disturbances. Hence, the structure of the composite controller includes the adaptive SMC for the slow subsystem for chaotic desired trajectory tracking and backstepping controller for the tip deflection suppression.

5.2 Chapter objectives

The following are the objectives of this chapter:

- To obtain a nominal model of the system from the modelling method by segregating the dynamics into a slow and fast subsystem.

- To design a sliding mode control using a linear matrix inequality for the sliding mode constants for the slow subsystem of the first composite controller.

- To design a state feedback controller using a linear matrix inequality for the fast subsystem of the first composite controller.

- To design an adaptive sliding mode control for the slow subsystem of the second composite controller.
- To design a backstepping controller for the fast subsystem of the second composite controller.
- To add the payload and the disturbances to the designed second composite controller tracking the chaotic desired trajectory.
- To compare the performances of the designed controllers with the earlier reported similar works.

The next section presents the modelling of TLFM using the assumed modes method by the singular perturbation technique.

5.3 Model decomposition by the singular perturbation technique

The control of the flexible manipulator is difficult as the system is underactuated in which all the modes of a link are to be controlled by adjusting a single actuating torque. The solution to this control problem involving underactuation has previously been achieved using singular perturbation [1]. It uses a perturbation parameter to segregate the complex dynamic systems into simpler subsystems of different time scales which have been applied successfully for controlling the flexible manipulators. The dynamic equation of a TLFM based on Lagrange's assumed modes method [1] is given in (6.19) and then simplified in (6.20). The dynamics (6.20) can be rewritten as

$$B\begin{pmatrix} \ddot{\theta} \\ \ddot{\eta} \end{pmatrix} + \begin{pmatrix} h_s(\theta,\dot{\theta},\eta,\dot{\eta}) \\ h_f(\theta,\dot{\theta},\eta,\dot{\eta}) \end{pmatrix} + \begin{pmatrix} D_1\dot{\theta} \\ D_2\dot{\eta} \end{pmatrix} + \begin{pmatrix} K_1\theta \\ K_2\eta \end{pmatrix} = \begin{pmatrix} I \\ 0 \end{pmatrix}\tau \tag{5.1}$$

where $\theta = [\theta_1\ \theta_2]^T$ is the joint angle, $\eta = [\eta_1^T\ \eta_2^T]^T$, $\eta_i = [\eta_{i1}\ \eta_{i2}]^T$ are the modes of the ith link and $i = 1, 2$. The mass matrix is a positive definite and its inverse is written as

$$B^{-1} = N = \begin{pmatrix} N_{11} & N_{12} \\ N_{21} & N_{22} \end{pmatrix} = \begin{pmatrix} M_s & M_{sf}^T \\ M_{sf} & M_f \end{pmatrix}^{-1} \tag{5.2}$$

where $M_s = (N_{11} - N_{12}N_{22}^{-1}N_{21})^{-1}$, $N_{11} \in R^{2X2}$, $N_{12} \in R^{2X4}$, $N_{21} \in R^{4X2}$, $N_{22} \in R^{4X4}$ and $K_1 = 0$. We can rewrite (5.1) as

$$\ddot{\theta} = -N_{11}(D_1\dot{\theta} + h_s + K_1\theta) - N_{12}(D_2\dot{\eta} + h_f + K_2\eta) + N_{11}\tau \tag{5.3}$$

$$\ddot{\eta} = -N_{21}(D_1\dot{\theta} + h_s + K_1\theta) - N_{22}(D_2\dot{\eta} + h_f + K_2\eta) + N_{21}\tau \tag{5.4}$$

For a singular perturbation, we define new state variables $\eta = \epsilon\delta$ and $K_s = \epsilon K_2$ where, ϵ is the perturbation parameter. The term $\frac{1}{\epsilon}$ is the smallest value of the stiffness constant matrix K. The singularly perturbed model of a TLFM, given in (5.3) and (5.4), can be described as

$$\ddot{\theta} = -N_{11}(D_1\dot{\theta} + h_s) - N_{12}(D_2\epsilon\dot{\delta} + h_f + K_s\delta) + N_{11}\tau \tag{5.5}$$

$$\ddot{\eta} = -N_{21}(D_1\dot{\theta} + h_s) - N_{22}(D_2\epsilon\dot{\delta} + h_f + K_s\delta) + N_{21}\tau \tag{5.6}$$

With the composite control strategy, the total torque can be divided as

$$\tau = \tau_s + \tau_f \tag{5.7}$$

where τ_s and τ_f are the torques for the slow and fast subsystems, respectively.

For identifying the slow subsystem, we set $\epsilon = 0$ in (5.6) and we get

$$\bar{\delta} = -K_s^{-1}\bar{N}_{22}^{-1}(\bar{N}_{21}\bar{D}_1\dot{\bar{\theta}} + \bar{N}_{21}\bar{h}_s + \bar{N}_{22}\bar{h}_f - \bar{N}_{21}\tau_s) \tag{5.8}$$

where the over bar indicates the value of the variable when $\epsilon = 0$.

Applying the two-time scale perturbation technique, the slow and the fast subsystems can be obtained. The slow subsystem is described as

$$\ddot{\bar{\theta}} = (\bar{N}_{11} - \bar{N}_{12}\bar{N}_{22}^{-1}\bar{N}_{21})(-\bar{D}_1\dot{\bar{\theta}} - \bar{h}_s + \tau_s) \tag{5.9}$$

Using (5.2), the slow subsystem can be represented as

$$\ddot{\bar{\theta}} = \bar{M}_s^{-1}(-\bar{D}_1\dot{\bar{\theta}} - \bar{h}_s + \tau_s) \tag{5.10}$$

In order to derive a boundary layer correction, a fast time scale, $\gamma = \frac{t}{\sqrt{\epsilon}}$ and boundary layer correction terms $x_1 = \delta - \bar{\delta}$ and $x_2 = \sqrt{\epsilon}\delta$ are defined.

The fast subsystem is described as

$$\dot{x}_f = A_f x_f + B_f \tau_f \tag{5.11}$$

where $A_f = \begin{pmatrix} 0 & I \\ -\bar{N}_{22}K_s & 0 \end{pmatrix}$; $B_f = \begin{pmatrix} 0 \\ \bar{N}_{21} \end{pmatrix}$; $x_f = \begin{pmatrix} x_1 \\ x_2 \end{pmatrix}$

The dynamics of the slow subsystem of a TLFM, given in (5.10), can be written in state space form as

$$\dot{y} = A(y)y + B(y)u \tag{5.12}$$

where $y = [\bar{\theta}^T \, \dot{\bar{\theta}}^T] = [y_1^T \, y_2^T]^T$, $y_1 = \bar{\theta}$, $y_2 = \dot{\bar{\theta}}$, $u = \tau_s$.

$$A(y) = \begin{pmatrix} 0 & I \\ 0 & -M_s^{-1}D_1 \end{pmatrix}, B(y) = \begin{pmatrix} 0 \\ -M_s^{-1} \end{pmatrix} \tag{5.13}$$

Here, the system matrix $A(y)$ and input matrix $B(y)$ are written as

$$A(y) = \begin{pmatrix} A_{11} & A_{12} \\ A_{21} & A_{22} \end{pmatrix}, B(y) = \begin{pmatrix} 0 \\ B_2 \end{pmatrix} \tag{5.14}$$

where $A_{11} = 0_{2X2}$, $A_{12} = I_{2X2}$, $A_{21} = 0_{2X2}$, $A_{22} = -M_s^{-1}D_1 \in R^{2X2}$, $B_2 = -M_s^{-1} \in R^{2X2}$.

Now, the state space model of (5.12) can be written as

$$\dot{y}_1 = A_{11}y_1 + A_{12}y_2 \tag{5.15}$$

$$\dot{y}_2 = A_{21}y_1 + A_{22}y_2 + B_2u \tag{5.16}$$

The next section describes the designing of the composite controller for the slow and the fast subsystems defined in (5.10) and (5.11).

5.4 Design of an LMI based SMC for the slow subsystem and LMI based state feedback controller (SFC) for the fast subsystem

In this section, a composite controller is designed for a TLFM using the dynamics designed in the above section. It consists of a LMI based SMC designed for the slow subsystem for tracking the tip trajectory and a LMI based state feedback for the fast subsystem for the suppression of the tip deflections. Here, a LMI is used for finding the optimised values of the gains of the controllers. The subsequent subsections discuss the design of the proposed composite controller.

5.4.1 Design of a LMI based SMC for the slow subsystem

Here, a normal SMC using LMI is designed for the trajectory tracking of the slow subsystem of a TLFM. Let $y_{di} \in R^{2X1}$ be the desired trajectories, then trajectories tracking errors dynamics are defined as

$$\begin{cases} e_1 = y_1 - y_{d1} \\ e_2 = y_2 - y_{d2} \end{cases} \tag{5.17}$$

where $y_{d1} = [\theta_{1d} \, \theta_{2d}]^T$, $y_{d2} = [\dot{\theta}_{1d} \, \dot{\theta}_{2d}]^T$.

Assumption 1. Consider y_{di} is consistent with the plant system matrix (5.14). Let the dynamics of the desired trajectories y_{di} be continuous and excited by u_y. It is described as:

$$\dot{y}_{d1} = A_{11}y_{d1} + A_{12}y_{d2} \tag{5.18}$$

$$\dot{y}_{d2} = A_{21}y_{d1} + A_{22}y_{d2} + u_y \tag{5.19}$$

Now, the normal sliding surface is defined as

$$s(e) = e_2 + \lambda e_1 \tag{5.20}$$

where $\lambda > 0$ is a constant matrix which is determined using LMI.

The necessary condition for the existence of the sliding condition is $s(e) = 0$. The equivalent sliding mode dynamics are described as

$$e_2 = -\lambda e_1 \tag{5.21}$$

Using (5.14)–(5.15), (5.18) and (5.19), the equivalent sliding mode dynamics for e_1 are described as

$$\dot{e}_1 = (A_{11} - A_{12}\lambda)e_1 \tag{5.22}$$

Theorem 5.1. *Suppose, there exist matrices* $\alpha,\ X,\ Y > 0$ *with appropriate dimensions and satisfy the following LMI:*

$$Q = \begin{pmatrix} A_{11}X - A_{12}Y + XA_{11}^T - Y^TA_{12}^T & X \\ X & -\alpha \end{pmatrix} < 0 \tag{5.23}$$

$$\Omega I_{n-m} - W > 0 \tag{5.24}$$

Then, the error dynamics (5.22) are asymptotically stable. The value of λ *can be obtained as* $\lambda = YX^{-1}$.

Proof. Considering a Lyapunov function candidate as

$$\begin{cases} V_1(e_1) = e_1^T R e_1 \\ \dot{V}_1(e_1) = e_1^T R\dot{e}_1 + \dot{e}_1^T R e_1 \end{cases} \tag{5.25}$$

where R a is positive definite matrix. Using (5.22), (5.25) can be written as

$$\dot{V}_1(e_1) = e_1^T[R(A_{11} - A_{12}\lambda) + (A_{11} - A_{12}\lambda)^T R]e_1 \tag{5.26}$$

Suppose the following inequality holds

$$R(A_{11} - A_{12}\lambda) + (A_{11} - A_{12}\lambda)^T R \leq -\alpha^{-1} \tag{5.27}$$

then

$$\begin{aligned}\dot{V}_1 &\leq e_1^T(-\alpha^{-1})e_1 \\ &\leq -(\mu_{min}(\alpha^{-1}))\|e_1\|^2\end{aligned} \tag{5.28}$$

where μ_{min} is the minimum eigenvalue of α^{-1}.

The Lyapunov function (5.28) can be simplified as

$$\dot{V}_1 \leq -\gamma V_1(e_1) \tag{5.29}$$

where,

$$\gamma = \frac{\mu_{min}(\alpha^{-1})}{\mu_{max}(R)} \tag{5.30}$$

For the positiveness of the parameter γ, the following can be obtained from (5.24) such that

$$\Omega > \mu_{max}(W) \tag{5.31}$$

Since, α and R are the positive definite matrices, the quantity γ is a positive scalar. Now, let $X = R^{-1}$, and pre and post multiplying X in (5.27) we get

$$A_{11}X - A_{12}\lambda X + XA_{11}^T - (\lambda X)^T A_{12}^T \leq -X\alpha^{-1}X \tag{5.32}$$

Considering $Y = \lambda X$ in (5.32) and using Shur complement [25], the LMI in (5.23) is satisfied. Thus, the equivalent sliding mode error dynamics (5.22) are asymptotically stable if the LMI given in (5.23) is feasible. □

The existence of the manipulator dynamics (5.12) on the sliding surface (5.20) is achieved using Theorem 5.1.

Theorem 5.2. *Consider, the sliding surface defined in (5.20) and suppose the value of λ is determined using LMI in (5.23), and if the control torque input for the slow subsystem of the TLFM dynamics (5.12) is defined as*

$$\tau_s = -M_s P_s - D_1 y_2 - \rho tanh(s) \tag{5.33}$$

where $P_s = -\lambda e_2 + \ddot{y}_{d2}$, 'tanh' is the tan hyperbolic function and ρ is a positive gain. Then, the trajectories of the flexible manipulator dynamics converge to the sliding surface (5.20) in finite time and remain on it.

Proof. Consider another Lyapunov function candidate as

$$V_2(s) = \frac{1}{2}ss^T \tag{5.34}$$

Taking the time derivative of (5.34) and using (5.12), it can be written using (5.20) as

$$\dot{V}_2 = \begin{cases} s^T\dot{s} \\ s^T(\dot{y}_2 - \dot{y}_{d2} + \lambda\dot{e}_1) \\ s^T(A_{21}y_1 + A_{22}y_2 + B_2\tau_s + d_l - P_s) \end{cases} \tag{5.35}$$

Now using the control torque defined in (5.33), (5.35) is written as

$$\dot{V}_2(s) \leq \sum_{i=1}^{2} |s_i|(\rho_i - l_l) \tag{5.36}$$

where $\rho_i > l_l$. Hence, the joint trajectories of the flexible manipulator (5.12) follow the desired trajectories using the control input torque proposed in (5.33). □

The next subsection describes the design of an LMI based state feedback controller for the fast subsystem to suppress the TLFM tip deflection.

5.4.2 Design of an LMI based SFC for the fast subsystem

In this subsection, a LMI based state feedback controller (SFC) is designed for regulating the dynamics of the fast subsystem of the TLFM to zero. Let (A_f, B_f) pairs in (5.11) be controllable, then a suitable state feedback controller can be designed as in (5.37).

$$\tau_f = -kx_f \tag{5.37}$$

where $k \in R^{2X8}$ is the positive definite matrix and its value is determined using the feasible solution of the LMI obtained from Theorem 5.3.

Theorem 5.3. *Suppose there exist matrices $P, S > 0$ with appropriate dimensions and the following LMI is satisfied:*

$$\begin{cases} P > 0 \\ A_fP - B_fS + PA_f^T - S^TB_f^T < 0 \end{cases} \tag{5.38}$$

Then, the fast subsystem is asymptotically stable along the equilibrium trajectory $\bar{\delta}$ defined in (5.8). The value of k in (5.37) can be obtained as $k = SP^{-1}$.

Proof. Suppose there exists a positive definite matrix P which satisfies the following inequality

$$A_cP + PA_c^T < 0 \tag{5.39}$$

where $A_c = (A_f - B_fk)$ is the closed loop system matrix of (5.11). The inequality (5.39) can be written as

$$(A_f - B_fk)P + P(A_f - B_fk)^T < 0 \tag{5.40}$$

Now, consider a matrix S which is described as $S = kP$, then (5.40) can be modified as

$$A_fP - B_fS + PA_f^T - S^TB_f^T < 0 \tag{5.41}$$

Thus, using (5.41), the LMI in (5.38) can be obtained. In (5.41), two matrices P and S are unknown whose value can be obtained from the feasible solution of the LMI in (5.38). □

Figure 5.1 shows the structure for the composite controller based on a two-time scale model of the manipulator given in (5.1). $\bar{\tau}$ is the slow control torque (5.33) and τ_f is the fast control torque (5.37). Boundary layer correction is applied to separate the two-time scales during the design process. Thus, the structure of the composite control technique using a LMI based SMC for slow subsystem and a LMI based state feedback control for the fast subsystem is shown in Fig. 5.1. In the next section, a new composite control is designed for the TLFM dynamics. An adaptive sliding mode control is designed for the slow subsystem and backstepping control is designed for the fast subsystem.

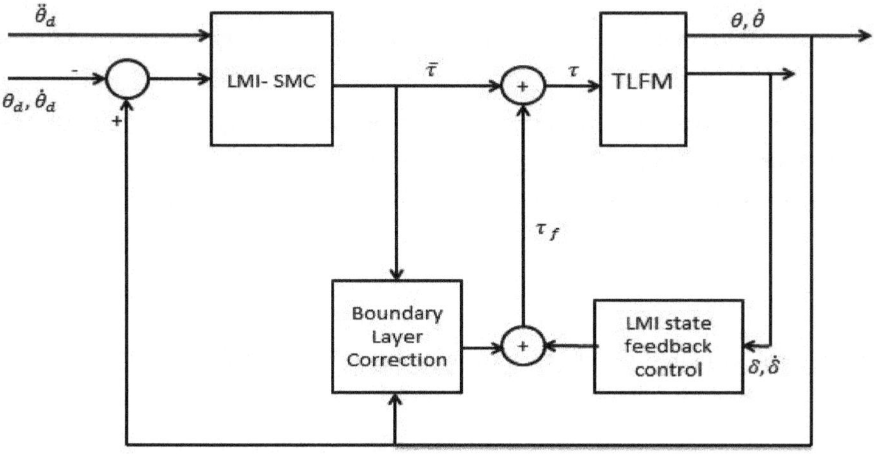

Figure 5.1: Structure of the composite control technique using a LMI based SMC and a LMI based state feedback controller.

5.5 Design of an adaptive SMC for the slow subsystem and backstepping controller of the fast subsystem

In the previous section, a composite control was designed using an LMI based SMC and state feedback control. The gains for the SMC and SFC were determined using Linear Matrix Inequalities (LMI). In this section, tracking of a desired chaotic signal and suppression of link deflection under the presence of variable payloads and bounded matched external disturbances are considered as problems. Hence, designing a robust controller is required to compensate for the disturbances and to suppress the tip deflection quickly. Therefore, a composite control is designed using an adaptive SMC and backstepping control for the mentioned problem. The consequent subsections describe the designing of the controllers for the slow and the fast subsystems.

5.5.1 Design of an adaptive SMC for the slow subsystem

This section describes the design of a composite control input $\tau_i = \tau_s + \tau_f$ for the TLFM dynamics (5.1). This is achieved by designing separate controllers τ_s using an adaptive SMC and τ_f using a backstepping control for the slow and the fast subsystems, respectively.

In order to track the chaotic desired trajectory by the TLFM dynamics (5.1) in the presence of bounded matched disturbances, an adaptive sliding mode control (A-SMC) is designed. The controller is designed for the slow subsystem of the TLFM (5.10).

The designing of A-SMC is achieved in two steps. The first step is the design of a suitable sliding surface and the second step is the determination of the sliding mode control law.

Suppose the dynamics of the slow subsystem of the TLFM is affected by the bounded and matched disturbances $\Delta f(\theta)$. Then, (5.10) can be expressed as

$$\ddot{\bar{\theta}} = -(\bar{B}_r)^{-1}(-\bar{H}_r - \bar{D}_r\dot{\bar{\theta}} + \tau_s) + \Delta f(\theta) \tag{5.42}$$

It is assumed that the system uncertainties $\Delta f(\theta)$ are bounded as $|\Delta f(\theta)| \leq L$. Let, θ_d be twice differentiable desired trajectories. Here, the desired signal θ_d is considered as a signal from a hidden chaotic attractors system. The tracking error for the slow subsystem is defined as

$$e_s = -\theta_d + \bar{\theta} \tag{5.43}$$

where $\bar{\theta}$ are the states of the slow subsystem. The sliding surface is defined as

$$s = \dot{e}_s + c_s e_s \tag{5.44}$$

where c_s is a positive definite constant gain matrix. When the system operates on the sliding mode, it satisfies the conditions $s = \dot{s} = 0$, i.e.,

$$\begin{cases} s(t) = \dot{e}_s + c_s e_s = 0 \\ \dot{s}(t) = \ddot{e}_s + c_s \dot{e}_s = 0 \end{cases} \tag{5.45}$$

Therefore, the equivalent sliding mode dynamics can be written as

$$\ddot{e}_s = -c_s \dot{e}_s, \; \dot{e}_{sy} = -c_s e_{sy}, \; assuming \; e_{sy} = -\dot{e}_s \tag{5.46}$$

Now, we show the stability of the equivalent sliding mode dynamics (5.46) using Lyapunov stability theory. A Lyapunov function candidate is selected as $V_{1s}(e_{sy}) = \frac{1}{2} e_{sy}^T e_{sy}$. The time derivative of the Lyapunov function candidate using (5.46) is written as

$$\dot{V}_{1s}(t) = \begin{cases} -e_{sy}^T c_s e_{sy} \\ -c_s \|e_{sy}\|^2 \end{cases} \tag{5.47}$$

Since, c_s is a positive definite matrix, $\dot{V}_{1s}$ is a negative definite. Thus, according to the Lyapunov stability theory, we can say that sliding motion on the sliding surface is stable and ensures the asymptotical convergence of the error dynamics to the origin.

After stabilising the equivalent sliding surface defined in (5.46), the next step is to design a sliding mode control law to drive the system trajectory onto the sliding mode, $s = 0$.

When the system is in the sliding mode, it satisfies $\dot{s} = 0$ and during the reaching phase [26], it satisfies

$$\dot{s} < -\rho \tanh(s) \tag{5.48}$$

Gain ρ is selected such that the reaching condition is satisfied and the sliding mode motion occurs. Now the control torque τ_s is designed to guarantee that the system trajectory hits $s = 0$. Using (5.42) and (5.45), the control torque τ_s can be obtained as

$$\tau_s = (\bar{B}_r)[-c_s \dot{e}_s + \ddot{\theta}_d + (\bar{B}_r)^{-1}(\bar{H}_r + \bar{D}_r \dot{\bar{\theta}}) - \rho \tanh(s)] \tag{5.49}$$

The ability of the control torque τ_s defined in (5.49) to drive the system (5.42) to the sliding mode $s = 0$, can be derived using Theorem 5.4.

Theorem 5.4. *Consider the uncertain slow system dynamics of the TLFM given in (5.42) controlled by τ_s in (5.49). Then, the trajectory of the slow subsystem (5.42) converges to the sliding surface $s = 0$.*

Proof. Let another Lyapunov function candidate be $V(s) = \frac{1}{2}s^T s$, then its time derivative can be written as

$$\dot{V}(s) = \begin{cases} s^T \dot{s} \\ s^T(\ddot{e}_s + c_s \dot{e}_s) \end{cases} \tag{5.50}$$

Now, using (5.42) and (5.49), (5.50) is written as

$$\dot{V}(s) = s^T\{(\bar{B}_r)^{-1}(-\bar{H}_r - \bar{D}_r\dot{\theta} + \tau_s) + \Delta f(\theta) - \ddot{\theta}_d + c_s\dot{e}_s\} \tag{5.51}$$

Using control input (5.49), (5.51) can be written as

$$\dot{V}(s) = s^T\dot{s} = s^T\{-\rho\tanh(s) + \Delta f(\theta)\} \tag{5.52}$$

Now, using the boundness of the uncertainties $|\Delta f(\theta)| \leq L$, (5.52) can be written as,

$$\dot{V}(s) \leq -\sum_{i=1}^{2} \rho_i|s_i| + L_i|s_i| \leq -\sum_{i=1}^{2} |s_i|(\rho_i - L_i) \tag{5.53}$$

With a suitable choice of $\rho_i > L_i$, we can ensure the negative definiteness of the Lyapunov function candidate $V(s)$. Therefore, the closed loop system (slow subsystem of the TLFM (5.10)) is asymptotically stable.

Now, it is assumed that the system uncertainties $\Delta f(\theta)$ are unknown and satisfy the condition $|\Delta f(\theta)| \leq L_i < \rho_i$. In real-life, the upper bound of the uncertainties is unknown and challenging to determine. In such a situation, the control law (5.49) is modified as:

$$\tau_s = (\bar{B}_r)[-c_s\dot{e}_s + \ddot{\theta}_d + (\bar{B}_r)^{-1}(\bar{H}_r + \bar{D}_r\dot{\theta}) - \hat{\rho}\tanh(s)] \tag{5.54}$$

where $\hat{\rho}$ is the estimate of ρ. To calculate the parameter $\hat{\rho}$, the following adaptation law is defined

$$\dot{\hat{\rho}} = k_\rho^{-1}|s| \tag{5.55}$$

where $k_\rho > 0$ is the adaptation gain matrix. Considering $\tilde{\rho} = \hat{\rho} - \rho$, the above adaptation law can be written as

$$\dot{\tilde{\rho}} = \dot{\hat{\rho}} = k_\rho^{-1}|s| \tag{5.56}$$

The stability of the adaptation law of the parameter $\hat{\rho}$ can be proved by considering a Lyapunov function candidate as

$$V_{2s}(s, \rho) = \frac{1}{2}(s^T s + k_\rho\tilde{\rho}^T\tilde{\rho}) \tag{5.57}$$

Taking time derivative of (5.57) and using (5.54) and (5.56), we can write it as

$$\dot{V}_{2s}(s,\rho) = s^T\dot{s} + k_\rho\tilde{\rho}^T\dot{\tilde{\rho}} = s^T\{-\hat{\rho}|s| + \Delta f(\theta)\} + \tilde{\rho}^T|s| \tag{5.58}$$

Using the boundness condition $|\Delta f(\theta)| \leq L_i < \rho_i$, we can write

$$\dot{V}(s) = -\sum_{i=1}^{2}\rho_i|s_i| + s^T\{\Delta f(\theta)\} = \sum_{i=1}^{2}L_i|s_i| - \rho_i|s_i| \leq -\sum_{i=1}^{2}(\rho_i - L_i)|s_i| \tag{5.59}$$

Since $\rho > 0$, we can say that (5.59) is a negative definite function. Therefore we can say that the trajectory of the slow subsystem dynamics (5.42) converges towards the sliding surface and remains on it. □

The next subsection describes the designing of backstepping control for the fast subsystem of the composite controller.

5.5.2 Design of a backstepping controller of the fast subsystem

In this subsection, a backstepping control technique is designed for controlling the fast subsystem. Dynamics of the fast subsystem (5.11) of the flexible manipulator can be written as:

$$\begin{cases} \dot{x}_1 = x_2 \\ \dot{x}_2 = -A_{f3}x_1 + B_{f2}\tau_f \end{cases} \tag{5.60}$$

where $A_{f3} = \bar{M}_{22}K_s$ and $B_{f2} = \bar{M}_{21}$. Suppose, x_d is a twice differentiable desired link deflection, and v is a virtual control variable. The link deflection errors are defined as

$$e_{1f} = x_d - x_1 \tag{5.61}$$

$$e_{2f} = v - x_2 \tag{5.62}$$

The errors dynamics can be obtained as

$$\dot{e}_{1f} = \dot{x}_d - x_2 \tag{5.63}$$

$$\dot{e}_{2f} = \dot{v} + A_{f3}x_1 - B_{f2}\tau_f \tag{5.64}$$

The control law designed for controlling the fast subsystem is obtained using Theorem 5.5.

Theorem 5.5. *Suppose the backstepping control law is defined in (5.65) using the error variable (5.63) and (5.64), then the fast subsystem of manipulator dynamics (5.60) follows the desired trajectory 'x_d', i.e., the link deflections of the manipulator are suppressed to zero properly.*

$$\tau_f = (B_{f2})^{-1}(\dot{v} + A_{f3}y_1 + k_2 e_{2f}) \tag{5.65}$$

Proof. The designing of a backstepping controller for the fast subsystem of TLFM (5.11) is achieved using the following steps:

Step 1: Let a Lyapunov function candidate be

$$V_{1f} = \frac{1}{2}e_{1f}^T e_{1f} \tag{5.66}$$

Time derivative of (5.66) using (5.63) and (5.62) is written as

$$\dot{V}_{1f} = e_{1f}^T(\dot{x}_d + e_{2f} - v) + e_{1f}e_{2f} \tag{5.67}$$

Now, let the virtual control variable v be

$$v = \dot{x}_d + k_1 e_{1f} + e_{2f} \tag{5.68}$$

where $k_1 > 0$ is a positive definite matrix. Using (5.68), the time derivative of Lyapunov function candidate (5.67) can be written as

$$\begin{cases} \dot{V}_{1f} & = -k_1 e_{1f}^T e_{1f} \\ & = -k_1 \|e_{1f}\|^2 \end{cases} \tag{5.69}$$

It is seen from (5.69) that the time derivative of Lyapunov function $\dot{V}_{1f}$ is negative definite. Thus, the first state variable of the fast subsystem (5.60) is stabilised. The next step is to demonstrate the stability of the second state variable and to calculate the control input τ_f in (5.65) for the fast subsystem.

Step 2: Let another Lyapunov function candidate be

$$V_{2f} = V_{1f} + \frac{1}{2}e_{2f}^T e_{2f} \tag{5.70}$$

Using (5.64) and (5.69) in the time derivative of (5.70), and it can be written as

$$\dot{V}_{2f} = -k_1\|e_{1f}\|^2 + e_2^T(\dot{v} + A_{f3}x_1 - B_{f2}\tau_f) \tag{5.71}$$

Now we can obtain the actual torque input as

$$\tau_f = (B_{f2})^{-1}(\dot{\upsilon} + A_{f3}x_1 + k_2 e_{2f}) \tag{5.72}$$

Using (5.72), $\dot{V}_{2f}$ in (5.71) is written as

$$\begin{cases} \dot{V}_{2f} &= -k_1 e_{1f}^T e_{1f} - k_2 e_{2f}^T e_{2f} \\ &= -k_1 \|e_{1f}\|^2 - k_2 \|e_{2f}\|^2 \end{cases} \tag{5.73}$$

Since k_1, k_2 are the positive definite constant matrices, then using the Lyapunov stability theory, we can say that (5.73) is a negative definite function. Thus, the error variables e_{1f} and e_{2f}, asymptotically converge to the origin with a suitable choice of constant matrices k_1, k_2. Therefore, the link deflections of the fast subsystem of the TLFM are suppressed to their desired values, i.e., to the origin quickly. □

Figure 5.2 shows the structure of the composite control technique designed for the chaotic desired trajectory tracking and tip deflection suppression using A-SMC and backstepping control. $\bar{\tau}$ is the slow torque (5.49) and τ_f (5.65) the fast control. The A-SMC is designed for the slow control and backstepping is designed for the fast one. A boundary layer correction is done for the segregation of the two-time scale division. Finally, both the controllers merge to give a total control input.

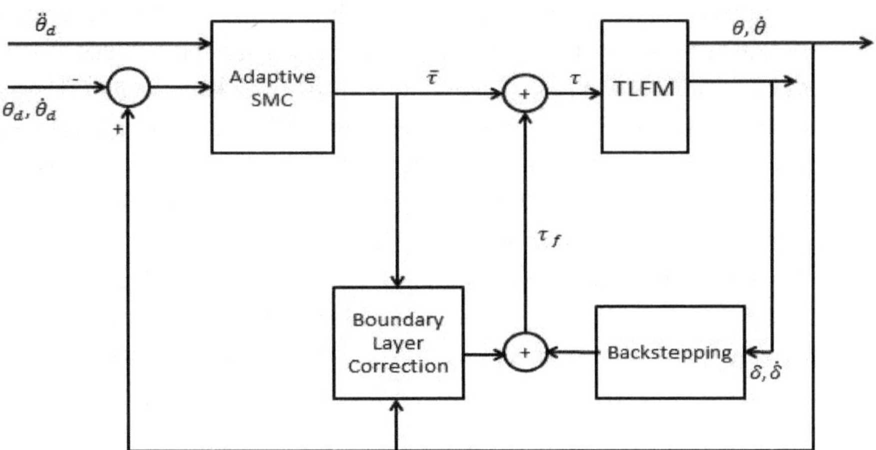

Figure 5.2: Structure of the composite control technique using an adaptive SMC and a backtepping controller.

5.6 Results and discussion

The results and discussion for the designed controllers of the two composite controllers are given in this section. The first subsection gives the results of the first composite controller which are discussed accordingly. The next subsection describes the chaotic desired signal used for the tracking control of the second composite control. The results are simulated for variable payloads in the subsequent subsections. Finally, the second proposed composite controller with a payload of 0.3 kg is compared with a controller already existing in the literature.

5.6.1 Results for the designed LMI based SMC for the slow subsystem and LMI based SFC for the fast subsystem

The desired trajectories for both the joints are considered to be:

$$\theta_{1d} = \frac{\pi}{4} - \frac{7}{6}e^{-\frac{3}{2}t} + \frac{7}{19}e^{-2t} \tag{5.74}$$

$$\theta_{2d} = \frac{\pi}{6} - \frac{7}{5}e^{-\frac{3}{2}t} + \frac{7}{11}e^{-2t} \tag{5.75}$$

The initial conditions for simulating the manipulator dynamics are given as $q = (0.05, 0, 0, 0, 0, 0, 0, 0)^T$. The values of different constants obtained from the LMI are given below:

$$\lambda = \begin{pmatrix} 11.8095 & 0 \\ 0 & 11.8095 \end{pmatrix}$$

$$k = \begin{pmatrix} -0.164 & 14.081 & 30.513 & 49.255 & 3.045 & -0.405 & 0.222 & -4.960 \\ 0.166 & -2.133 & -4.816 & -10.295 & -0.505 & -0.565 & -0.277 & 0.713 \end{pmatrix}$$

The other constants used for simulating the SMC of the slow subsystem are $\rho_1 = 10$, $\rho_2 = 10$. These parameter values are chosen by trial and error to maintain a good tracking performance and low control input. Figure 5.3 shows the joint trajectory tracking of links. It is seen that TLFM tracks the desired trajectories properly within a small time duration of $2s$ for both the links. The obtained two modes of each link are shown in Fig. 5.4. The tip deflection in Fig. 5.4 is the combination of the modes with its mode shapes. It is seen in Fig. 5.4 that the tip deflections of the first link and the second link are suppressed within 0.05 mm and 0.005 mm, respectively. The proposed sliding surfaces for the slow subsystem are shown in Fig. 5.6. Figures 5.7 and 5.8 show the proposed control inputs for the slow subsystem dynamics and fast subsystem dynamics, respectively. The composite control inputs which are a combination of slow and fast subsystems control inputs are shown in Fig. 5.9. It is observed from Fig. 5.9 that the maximum torque requirement for the first and second link are within bounds of $[-35, 10]$ Nm and $[-15, 1]$ Nm, respectively.

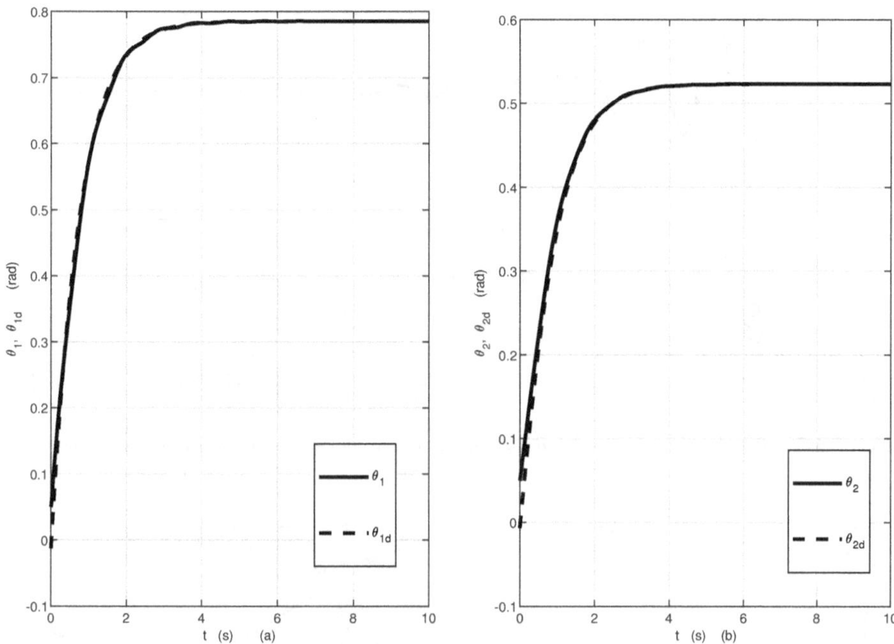

Figure 5.3: Joint trajectory tracking of: (a) link-1 and (b) link-2 using the composite controller in Section 5.4.

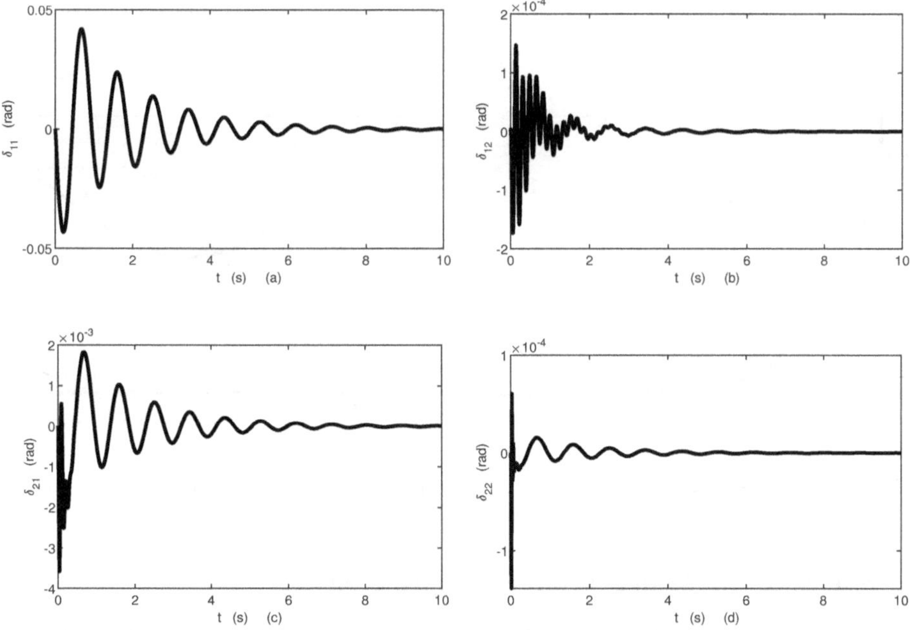

Figure 5.4: Modes of the respective link: (a) mode-1 of the link-1, (b) mode-2 of the link-1, (c) mode-1 of the link-2 and (d) mode-2 of the link-2 using the composite controller in Section 5.4.

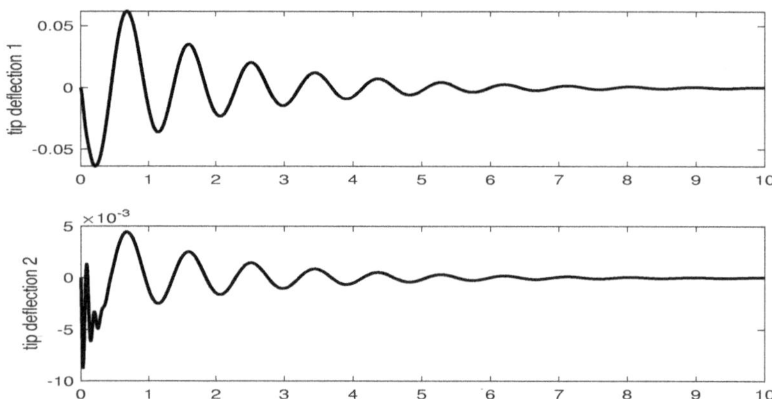

Figure 5.5: Tip deflections of: (a) link-1 and (b) link-2 using the composite control as discussed in Section 5.4.

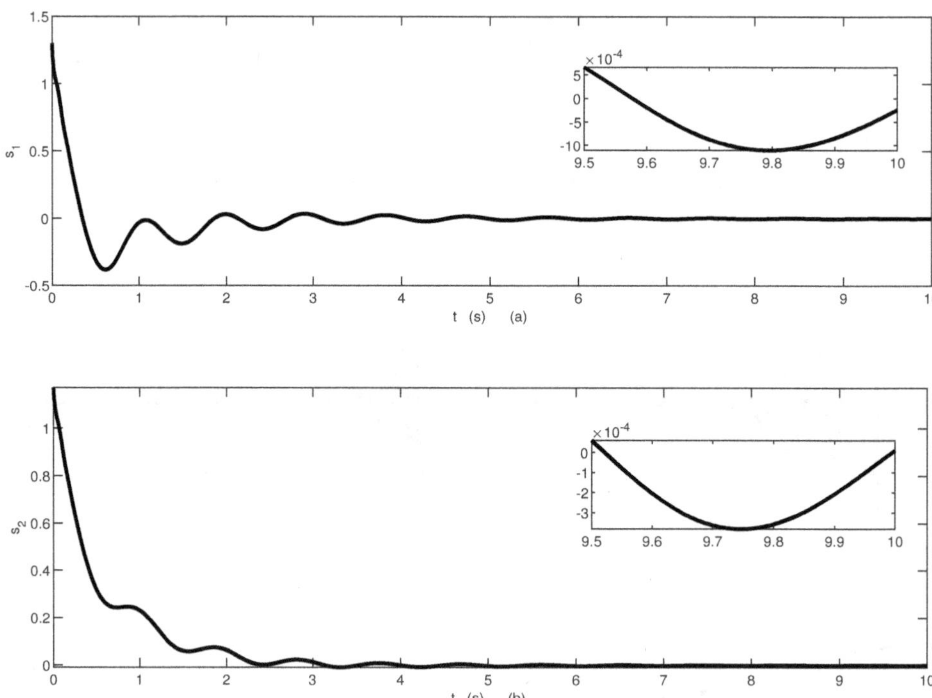

Figure 5.6: Sliding surfaces of the slow subsystem of composite control as discussed in Section 5.4: (a) link-1 and (b) link-2.

5.6.2 Hidden chaotic signals as the desired trajectories

The dynamics of a hidden attractors chaotic system whose signal is used as the desired trajectory for the TLFM is described as [27]

$$\begin{cases} \dot{y}_1 = y_2 \\ \dot{y}_2 = y_3 \\ \dot{y}_3 = -y_1 - 2.9y_3^2 + y_1y_2 + 1.1y_1y_3 - 1 \end{cases} \tag{5.76}$$

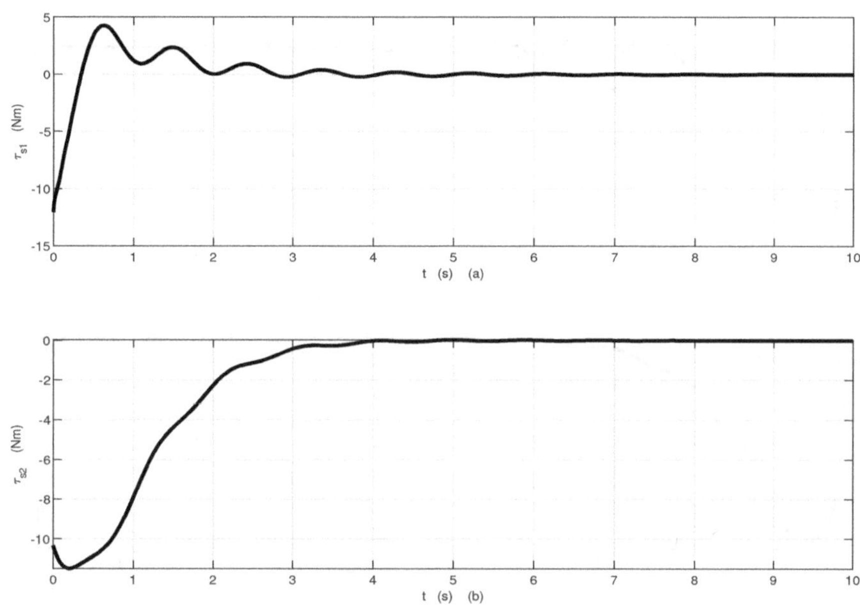

Figure 5.7: Required torques of the slow subsystem of composite control as discussed in Section 5.4: (a) link-1 and (b) link-2.

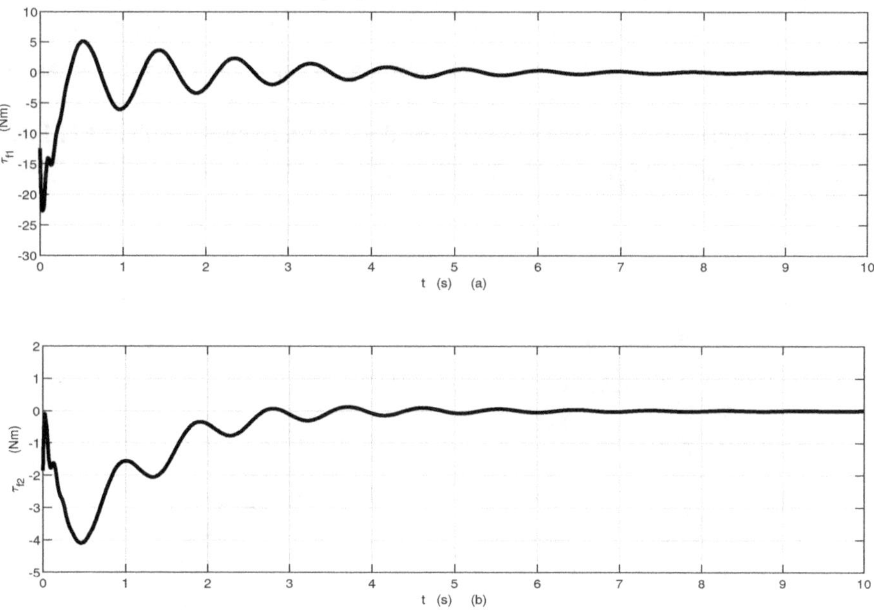

Figure 5.8: Required torques of the fast subsystem of composite control as discussed in Section 5.4: (a) link-1 and (b) link-2.

System (5.76) is chaotic with initial conditions $x(0) = (-2.2, 0.6, 0)^T$ where the Lyapunov exponents of the system are $L_i = (0.0638, 0, -1.0638)$. System (5.76) has equilibrium points at

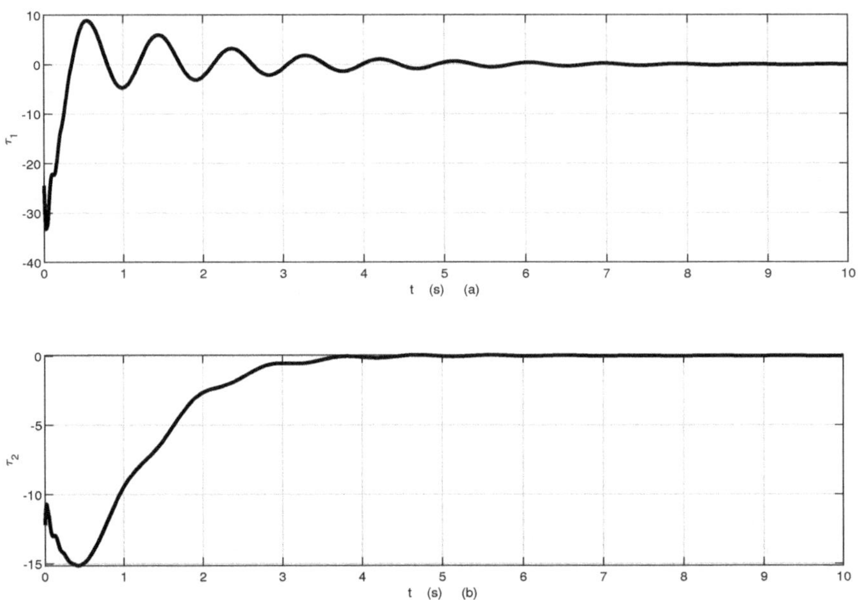

Figure 5.9: Composite control torque inputs of the flexible manipulator: (a) link-1 and (b) link-2 as discussed in Section 5.4.

$E = (-1, 0, 0)$ and stable eigenvalues corresponding to these equilibrium points. Thus, system (5.76) has hidden chaotic attractors [27]. The chaotic attractors and chaotic signals of system (5.76) with initial conditions $x(0) = (-2.2, 0.6, 0)^T$ are shown in Fig. 5.10 and Fig. 5.11, respectively.

In this chapter, the signals x_1, x_2, x_3 are used as the desired trajectories for the TLFM as $\theta_{d1} = \theta_{d2} = x_1$, $\dot{\theta}_{d1} = \dot{\theta}_{d2} = x_2$ and $\ddot{\theta}_{d1} = \ddot{\theta}_{d2} = x_3$.

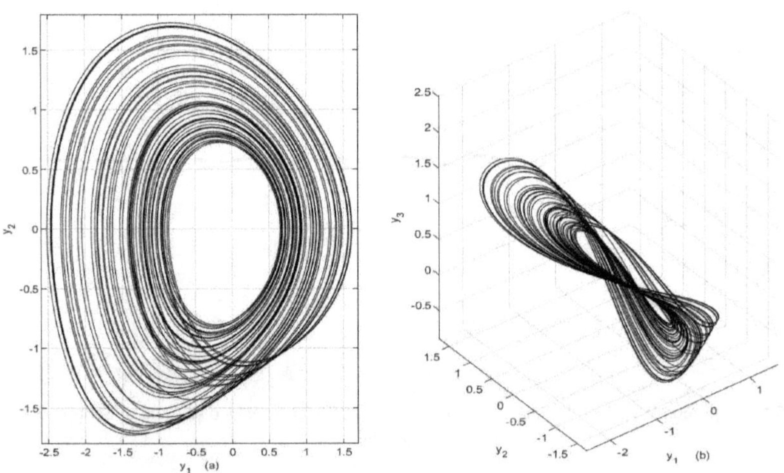

Figure 5.10: Hidden chaotic attractors of system (5.76).

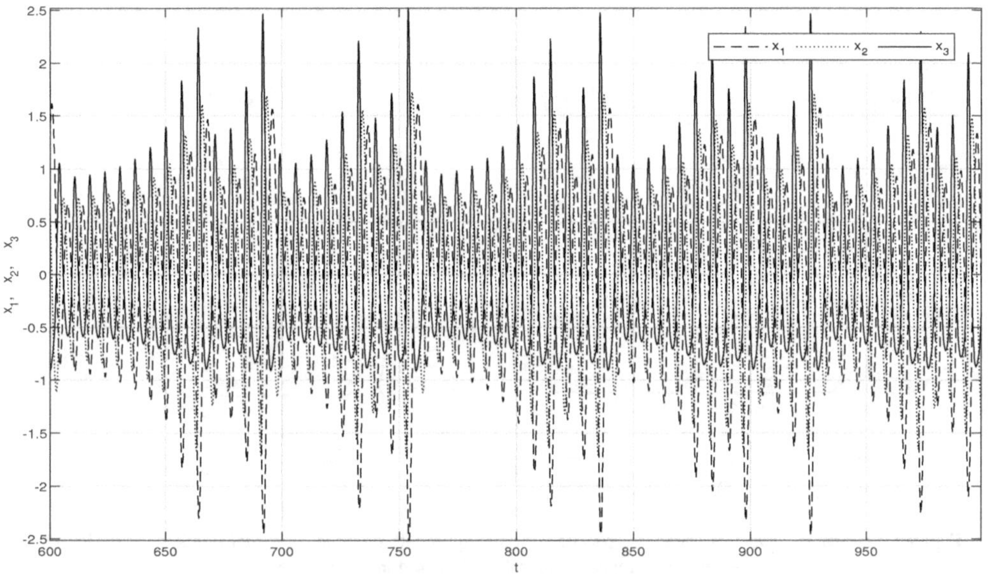

Figure 5.11: Hidden chaotic signals of system (5.76).

5.6.3 Results for the designed adaptive SMC for the slow subsystem and backstepping controller for the fast subsystem

Here, the initial conditions used for the simulation of a two-link flexible manipulator dynamics (5.1) and adaptation laws (5.55) are considered to be $(\theta(0), q(0) = (0.1, 0.1, 0.0, 0.0, 0.0, 0.0)^T$, $\dot{\theta}(0), \dot{q}(0) = (0, 0, 0, 0, 0, 0)^T$. The system uncertainties added in the manipulator dynamics are given as

$$\Delta f(\theta) = [0.1 \sin(\theta) \sin(\dot{\theta}), 0.1 \sin(\theta) \sin(\dot{\theta})]^T \tag{5.77}$$

The values of the sliding surface constant gains and adaptation constant gains used for adaptive SMC (A-SMC) (5.54) are $c_s = \begin{pmatrix} 20 & 0 \\ 0 & 20 \end{pmatrix}$ and $k = \begin{pmatrix} 14 & 0 \\ 0 & 14 \end{pmatrix}$, respectively, and the gains of the backstepping controller (5.65) are considered to be $k_1 = k_2 = 10$. These gains are considered in a manner to achieve better tracking performances using less control efforts.

5.6.4 Results with a nominal payload (0.145 kg)

Hidden chaotic desired trajectory tracking for the two-link flexible manipulator (5.1) is first discussed here with a nominal payload of $(0.145kg)$ and in the presence of system uncertainties (5.77). The trajectory tracking errors for both the links are shown in Fig. 5.12. It is observed from Fig. 5.12 that the chaotic trajectory tracking for both the links are achieved a time duration of $6\ s$. The modes of the first and second flexible links with a 0.145 kg payload are shown in Fig. 5.13 and Fig. 5.14, respectively. It is apparent from Figs. 5.13 and 5.14 that the modes of the links are suppressed within values $[10^{-3}, 10^{-5}]$ and $[10^{-4}, 10^{-6}]$ mm, respectively. The behaviour of the tip deflection for both

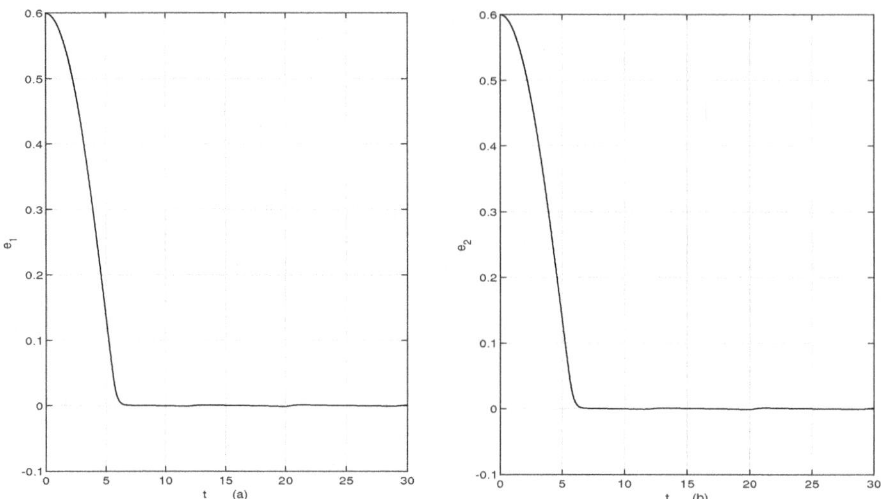

Figure 5.12: Chaotic trajectory tracking errors of the joint angles with nominal payload: (a) link-1 and (b) link-2 using the composite control as discussed in Section 5.5.

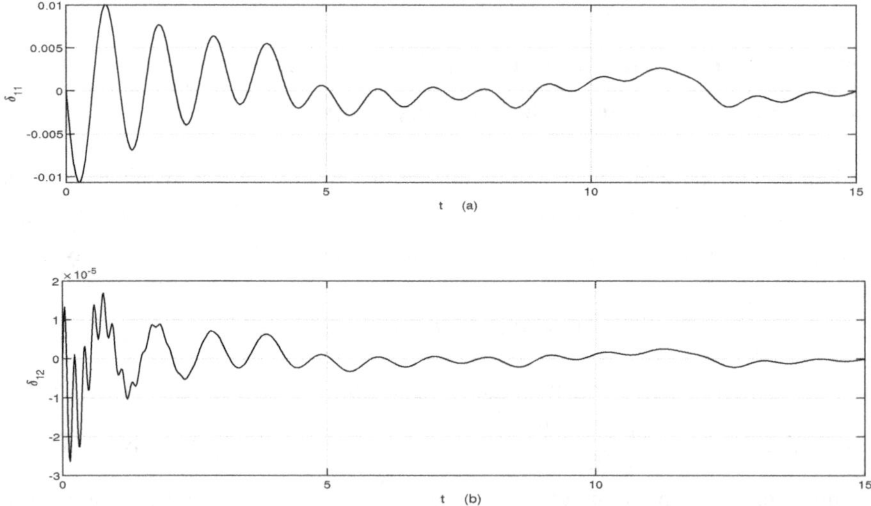

Figure 5.13: Modes of the link-1 with nominal payload using the composite control as discussed in Section 5.5.

the links are shown in Fig. 5.15. It is noted from Fig. 5.15 that the tip deflections of both the links are suppressed within the values 10^{-2} mm and 10^{-3} mm, respectively. Responses of the sliding surfaces designed for the slow subsystem of both the links are shown in Fig. 5.16. The required control torque inputs in the slow subsystem are shown in Fig. 5.17. It is seen from Figs. 5.16 and 5.17 that the chattering is very less in the sliding surfaces and control inputs. Responses of the control input required in the fast subsystem are shown in Fig. 5.18. Responses of the composite control input designed for the two-link flexible manipulator are shown in Fig. 5.19. It is noted from Fig. 5.19 that the required control torques using the composite control for both the links are initially high but after

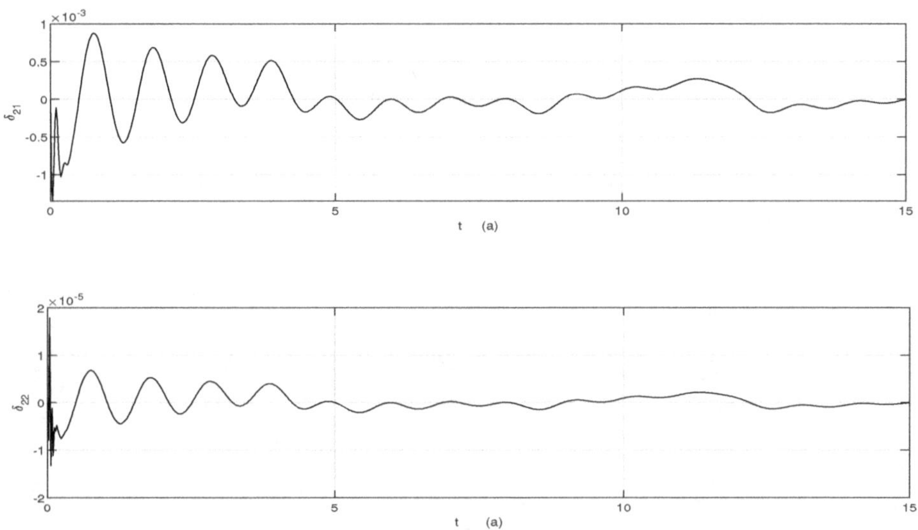

Figure 5.14: Modes of the link-2 with nominal payload using the composite control as discussed in Section 5.5.

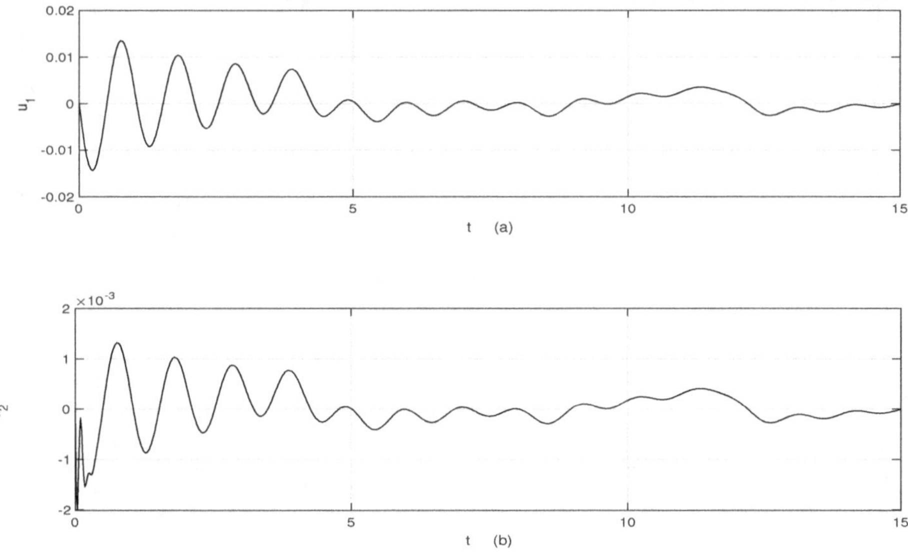

Figure 5.15: Responses of tip deflection of both the links under chaotic trajectory tracking with nominal payload using the composite control in Section 5.5.

sometime, these torques decay within small values. It is also seen from Fig. 5.19 that the variation in control inputs is low.

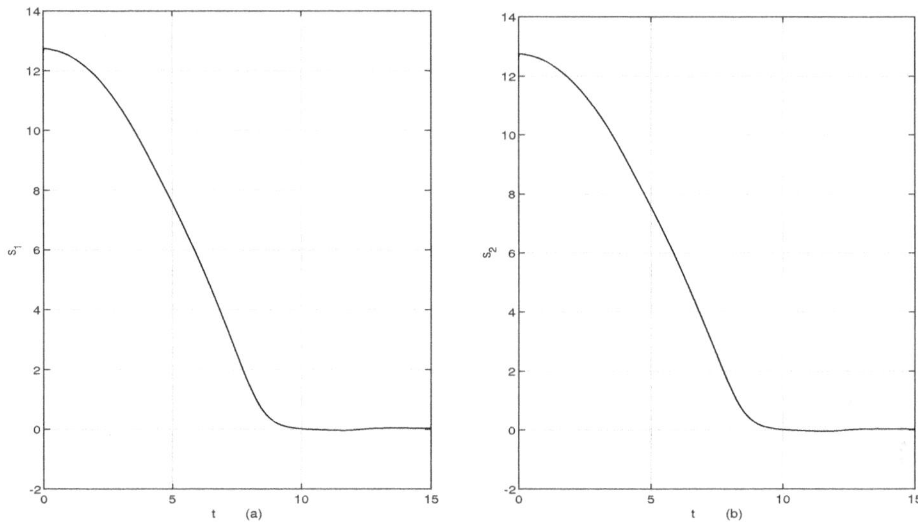

Figure 5.16: Sliding surfaces of the slow subsystem with nominal payload as discussed in Section 5.5.

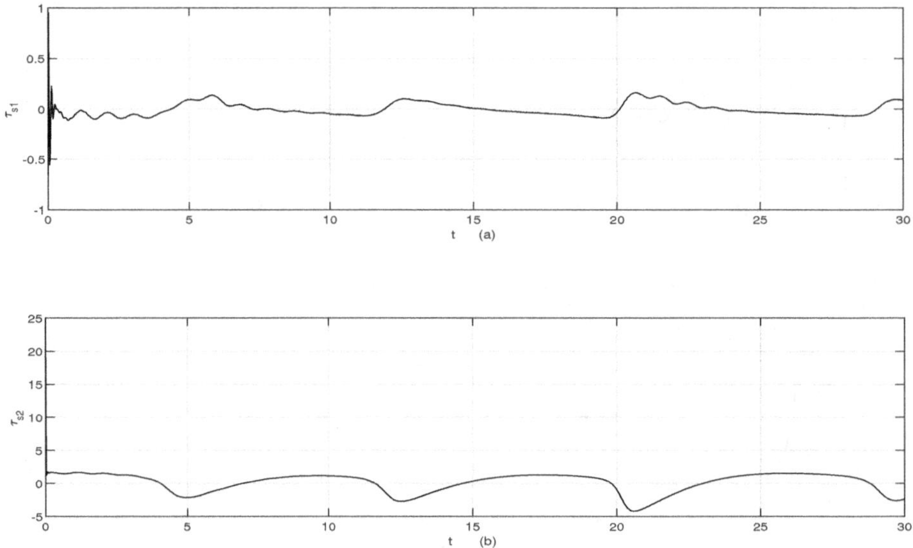

Figure 5.17: Required control torques during chaotic trajectory tracking of the slow subsystem of the composite control as discussed in Section 5.5 with the nominal payload.

5.6.5 Results with a 0.3 kg payload

This subsection describes the robustness of the controllers for chaotic tracking results of a two-link flexible manipulator (5.1) with a payload of 0.3 kg. The chaotic trajectory tracking errors for both the links with a 0.3 kg payload are shown in Fig. 5.20. It is apparent from Fig. 5.20 that with the addition of payload, the settling time (8 s) of the tracking errors increases as compared with the

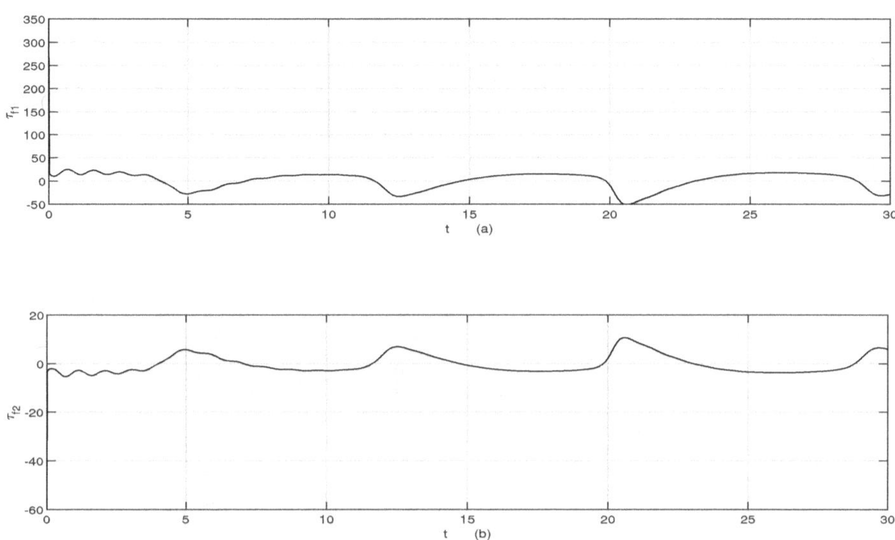

Figure 5.18: Required control torque for the fast subsystem with the nominal payload as discussed in Section 5.5.

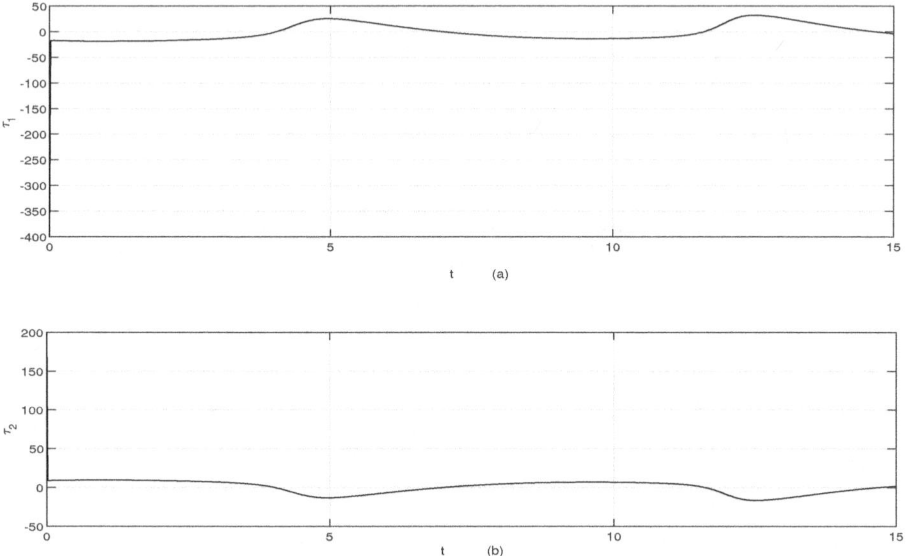

Figure 5.19: The composite control torques during chaotic trajectory tracking in Section 5.5 for the two-link flexible manipulator with the nominal payload.

nominal payload (6 s). Behaviour of modes of both the links with a 0.3 kg payload are shown in Figs. 5.21 and 5.22, respectively. The responses of the tip deflection of both the links are shown in Fig. 5.23. It is noted from Fig. 5.23 that with an increase in payload there is a small increase in the magnitudes of deflections for both the links. The behaviour of the sliding surfaces designed for the slow subsystem with a 0.3 kg payload are shown in Fig. 5.24. It is seen from Fig. 5.24 that there is an increase in the reaching time duration ($3\ s$) for the sliding surfaces with a $0.3\ kg$ payload

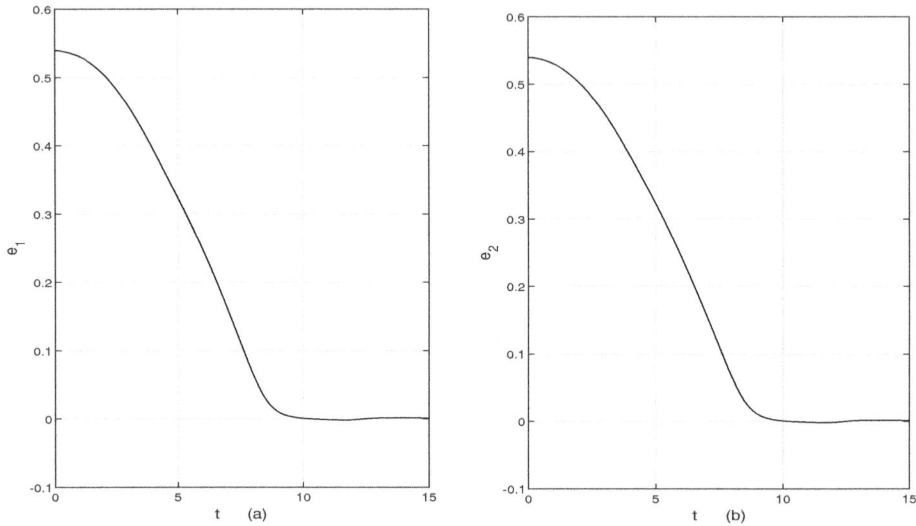

Figure 5.20: Chaotic trajectory tracking errors of the joint angles with 0.3 kg payload: (a) link-1 and (b) link-2 using the composite control in Section 5.5.

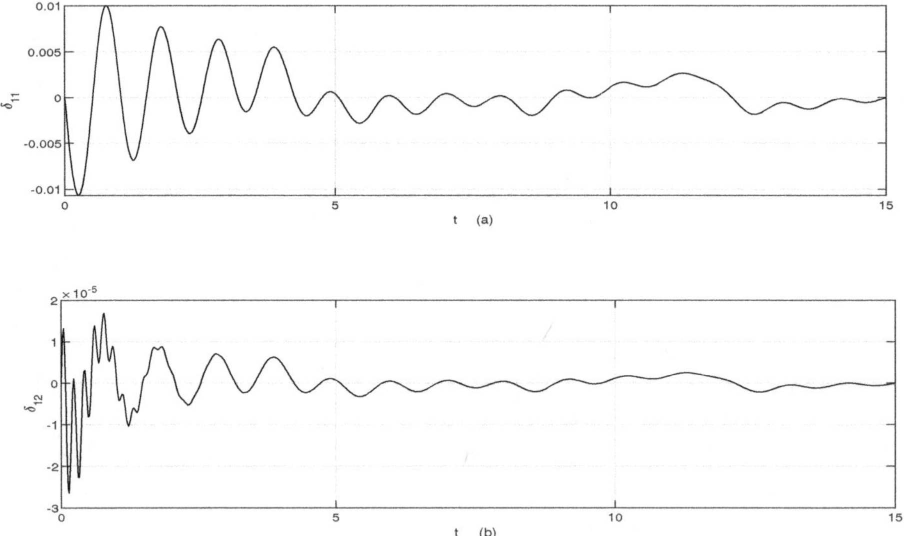

Figure 5.21: Modes of link-1 with 0.3 kg payload using the composite control in Section 5.5.

as compared with the nominal payload. Responses of the composite control inputs generated for the two-link flexible manipulator with a 0.3 kg payload are shown in Fig. 5.25. It is visible from Fig. 5.25 that due to an increase in the payload on the tip of the second link, it requires more control efforts as compared with the nominal payload.

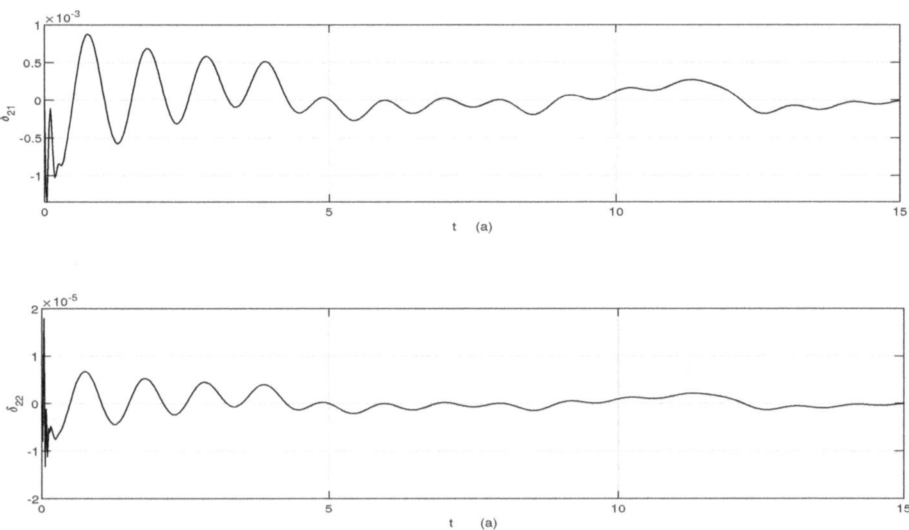

Figure 5.22: Modes of link-2 with 0.3 kg payload using the composite control in Section 5.5.

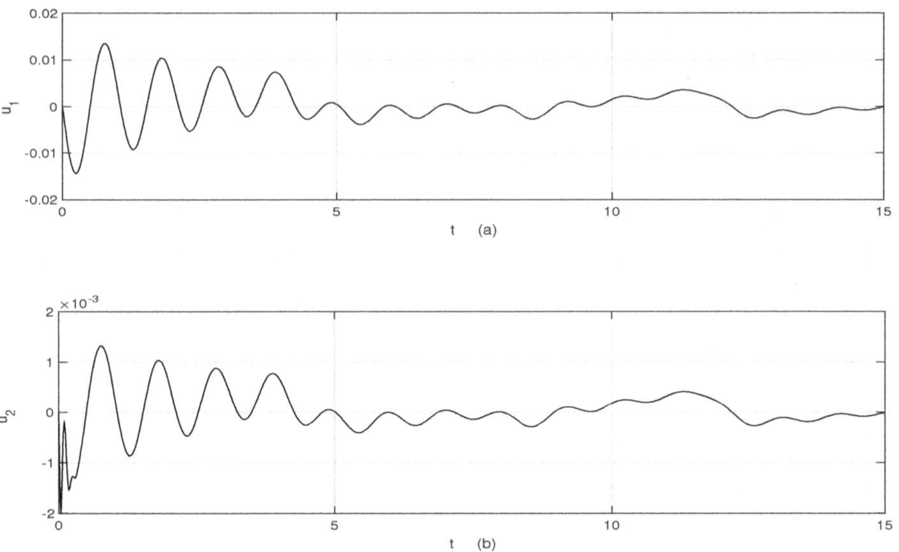

Figure 5.23: Responses of the tip deflections of both the link under chaotic trajectory tracking with 0.3 kg payload using the composite control in Section 5.5.

5.6.6 Comparison of the proposed composite controller (discussed in Section 5.5) with the controller referred to in [14]

Chaotic trajectory tracking performances of the proposed composite controller with a nominal payload for TLFM (5.1) are compared with the composite controller available [14]. Tracking errors using the proposed composite controller discussed in section 5.5 and the controller referred to in

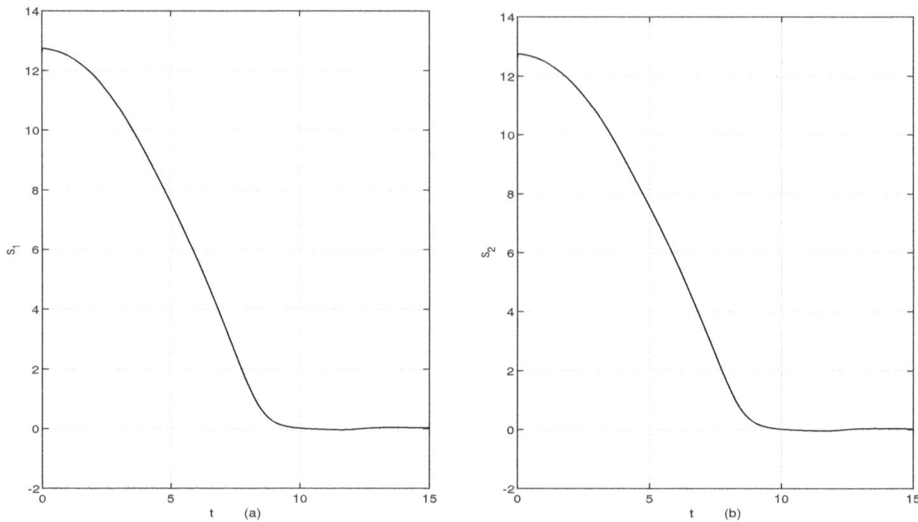

Figure 5.24: Sliding surfaces of the slow subsystem with 0.3 kg payload using the composite control discussed in Section 5.5 with 0.3 kg payload.

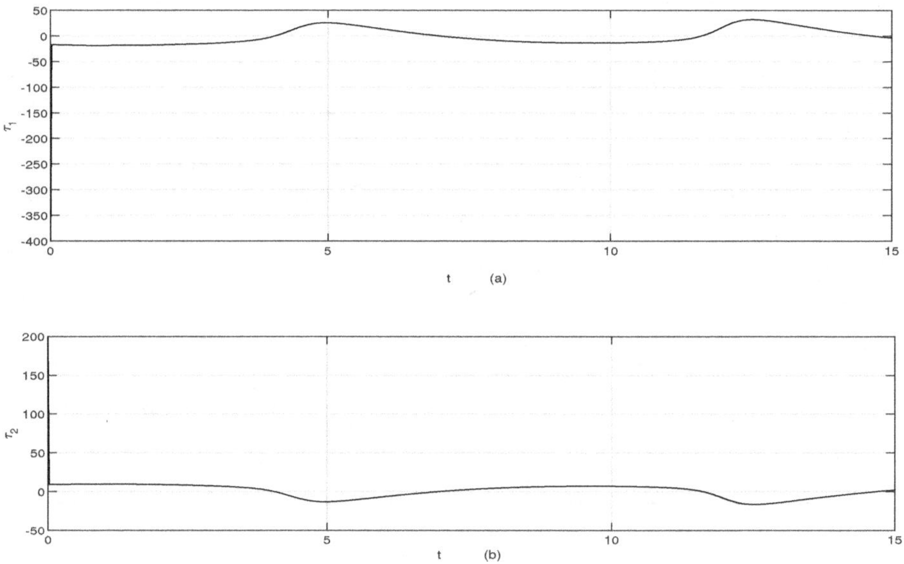

Figure 5.25: The composite control torques in Section 5.5 for the two-link flexible manipulator with 0.3 kg payload (during chaotic trajectory tracking).

[14] for both the links are shown in Fig. 5.26. It is apparent from Fig. 5.26 that the settling time for both the controllers is almost the same. But, the steady state error using the proposed composite controller is lower as compared to the controller referred to in [14]. The behaviours of the tip deflections using both the composite controllers are shown in Fig. 5.27. It is observed from Fig. 5.27 that tip deflection using the proposed controller is lower as compared to the controller referred to

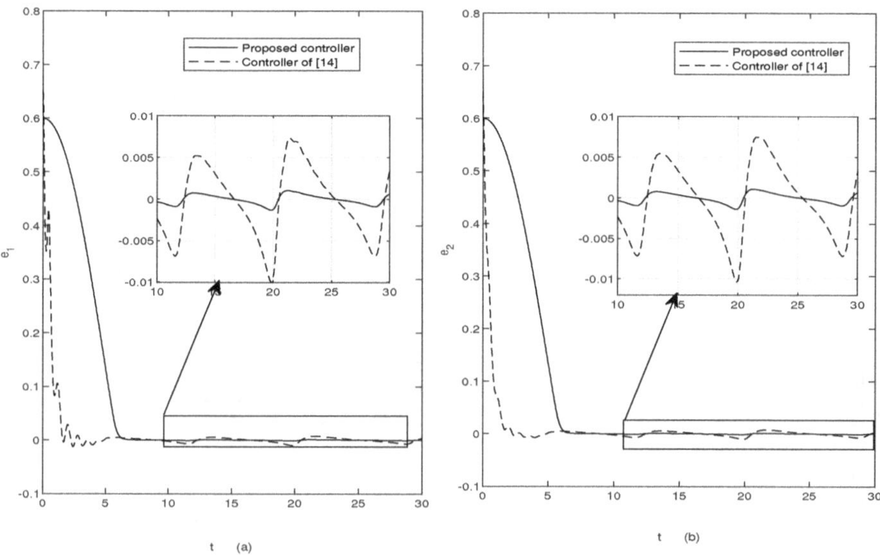

Figure 5.26: Chaotic trajectory tracking errors of the joint angles with nominal payload: (a) link-1 and (b) link-2 using the proposed composite controller as discussed in Section 5.5 and the controller of [14].

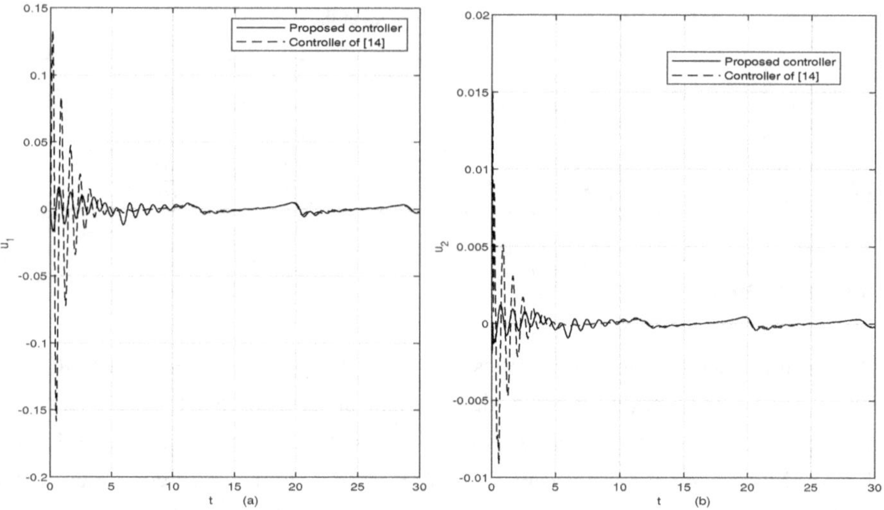

Figure 5.27: Tip deflections of both the links for chaotic path planning with nominal payload using the proposed composite controller as discussed in Section 5.5 and the controller of [14].

in [14]. The natures of the proposed composite control technique and composite controller referred to in [14] are shown in Fig. 5.28. It is noted from Fig. 5.28 that the required control efforts using the proposed composite controller is comparatively less than that of the controller [14] in link-2 and the same as in link-1. Thus, from the results of Figs. 5.26 to 5.28, we can say that the proposed composite controller performs better than the controller in [14] in terms of steady state tracking error, tip deflection and control efforts. Table 5.3 shows the comparison of the performances of the controller.

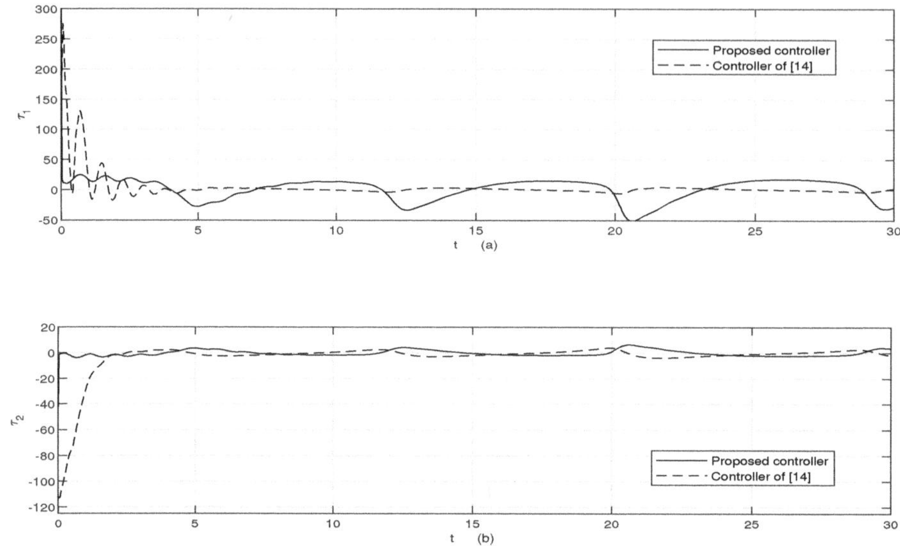

Figure 5.28: Comparison of the nature of the composite control torques in Section 5.5 and the controller of [14] for the two-link flexible manipulator with nominal payload during chaotic trajectory tracking.

Table 5.3: Comparison on performances of the controllers.

	Performances of controllers			
Control techniques	Proposed Controller		Controller in [14]	
Performance Indices	Link-1	Link-2	Link-1	Link-2
Control energy ($\|\|u\|\|_2$)	**98.0413**	**45.7030**	176.6112	555.5637
ISE of tip deflections	**0.058**	**0.0047**	0.164	0.011

5.7 Chapter summary

This chapter has presented the model decomposition of the dynamics of the system by the assumed modes method using a singular perturbation, designing of the two composite controllers and comparison of a composite controller with an existing controller. The first composite controller includes the LMI based SMC for the slow subsystem and LMI based state feedback controller for the fast subsystem. The second composite controller includes the adaptive SMC for the slow subsystem and backstepping controller for the fast subsystem for the tip trajectory tracking and suppression of tip deflection of a two-link flexible manipulator while it is subjected to handle variable payloads. Also, a chaotic system with a hidden chaotic attractor is used for the generation of the desired signals. The simulation performances of the two composite controllers have been obtained after implementing those control algorithms to the dynamic model of TLFM. The proposed second composite controller has smaller steady state errors, quick and smaller tip deflections and required lesser control efforts when compared with the controller referred to in (Mirzaee et al., 2010).

References

[1] B. Subudhi and A. S. Morris. Dynamic Modelling, Simulation and Control of A Manipulator with Flexible Links and Joints. *Robotics and Autonomous Systems*, vol. 41, no. 4, pp. 257–270, 2002.

[2] M. H. Korayem and M. Haghpanahi. Finite Element Method and Optimal Control Theory for Path Planning of Elastic Manipulators. *New Advances in Intelligent Decision Technologies*, vol. 199, no. 2009, pp. 117–126, 2009.

[3] B. Siciliano and W. J. Book. *A Singular Perturbation Approach to Control of Lightweight Flexible Manipulators*. PhD thesis, 1986.

[4] F. Khorrami, S. Jain, and A. Tzes. Experiments on Rigid Body-Based Controllers with Input Preshaping for a Two-Link Flexible Manipulator. *IEEE Transaction on Robotics and Automation*, vol. 10, no. 11, pp. 55–65, 1994.

[5] F. Khorrami and S. Jain. Non-Linear Control with End-Point Acceleration Feedback for a Two-Link Flexible Manipulator: Experimental Results. *Journal of Robotic Systems*, vol. 10, no. 4, pp. 505–530, 1993.

[6] X. U. Bo and Y. Bakakawa. Control Two-link Flexible Manipulators using Controlled Lagrangian Method. In *SICE Annual Conference in Sapporo*, pp. 289–294, 2004.

[7] S. H. Lee and C. W. Lee. Hybrid Control Scheme for Robust Tracking of Two-link Flexible Manipulator. *Journal of Intelligent & Robotic Systems*, vol. 34, no. 4, pp. 431–452, 2002.

[8] Y. Li, G. Liu, T. Hong, and K. Liu. Robust Control of A Two-link Flexible Manipulator with Quasi-static Deflection Compensation using Neural Networks. *Journal of Intelligent & Robotic Systems*, vol. 44, no. 3, pp. 263–276, 2005.

[9] Y. Zhang, Y. Mi, M. Zhu, and F. Lu. Adaptive Sliding Mode Control For Two-Link Flexible Manipulator with H infinity tracking. No. August, pp. 702–707, 2005.

[10] F. Matsuno and K. Yamamoto. Dynamic Hybrid Force/Position Control of Two Degrre of Freedom Flexible Manipulator. *Journal of Robotic Systems*, vol. 11, no. 5, pp. 355–366, 1994.

[11] Y. Wang, Y. Feng, and X. Yu. Fuzzy Terminal Sliding Mode Control of Two-link Flexible Manipulators, vol. 2, pp. 1620–1625, 2008.

[12] A. Ashayeri and M. Farid. Trajectory Tracking for Two-Link Flexible Arm via Two-Time Scale and Boundary Control Methods. In *Proceedings of IMECE2008 2008 ASME International Mechanical Engineering Congress and Exposition*, pp. 1–9, 2008.

[13] Y. Li, B. Tang, Z. Shi, and Y. Lu. Experimental Study for Trajectory Tracking of A Two-link Flexible Manipulator. *International Journal of Systems Science*, vol. 31, no. 1, pp. 3–9, 2010.

[14] E. Mirzaee, M. Eghtesad, and S. A. Fazelzadeh. Maneuver Control and Active Vibration Suppression of A Two-link Flexible Arm using A Hybrid Variable Structure/Lyapunov Control Design. *Acta Astronautica*, vol. 67, no. 9-10, pp. 1218–1232, 2010.

[15] D. Yue-jiao, C. Xi, Z. Ming, and R. Jun. Anti-windup for Two-link Flexible Arms with Actuator Saturation using Neural Network. In *2010 International Conference on E-Product E-Service and E-Entertainment (ICEEE),*, no. 06, pp. 6–9, 2010.

[16] Y. Wang, F. Han, Y. Feng, and X. Hongwei. Hybrid Continuous Nonsingular Terminal Sliding Mode Control of Uncertain Flexible Manipulators. In *40th IEEE Annual Conference of the Industrial Electronics Society (IECON)*, no. 51307035, pp. 190–196, 2014.

[17] S. Ding, J. Wang, and W. X. Zheng. Second-Order sliding mode control for nonlinear uncertain systems bounded by positive functions. *IEEE Transactions on Industrial Electronics*, vol. 62, no. 9, pp. 5899–5909, 2015.

[18] A. Nasiri, S. Kiong Nguang, and A. Swain. Adaptive sliding mode control for a class of MIMO nonlinear systems with uncertainties. *Journal of the Franklin Institute*, vol. 351, no. 4, pp. 2048–2061, 2014.

[19] H. Alwi and C. Edwards. Sliding mode fault-tolerant control of an octorotor using linear parameter varying-based schemes. *IET Control Theory & Applications*, vol. 9, no. 4, pp. 618–636, 2015.

[20] Y. Aoustin and A. Formal'sky. On the Feedforward Torques and Reference Trajectory for Flexible Two-Link Arm. *Multibody System Dynamics*, vol. 3, pp. 241–265, 1999.

[21] S. K. Pradhan and B. Subudhi. Nonlinear Adaptive Model Predictive Controller for a Flexible Manipulator: An Experimental Study. *IEEE Transactions on Control Systems Technology*, vol. 22, no. 5, pp. 1754–1768, 2014.

[22] M. Masoud, G. Mostafa, and S. N. Mostafa. Observer Based Tip Tracking Control of Two-link Flexible Manipulator. In *International Conference on Control, Automation and Systems 2010*, pp. 9–13, 2010.

[23] L. Zhang and J. Liu. Observer-based Partial Differential Equation Boundary Control for A Flexible Two-link Manipulator in Task Space. *IET Control Theory & Applications*, vol. 6, no. 13, pp. 2120–2133, 2012.

[24] S. K. Pradhan and B. Subudhi. Real-Time Adaptive Control of a Flexible Manipulator Using reinforcement learning. *IEEE Transactions on Automation Science and Engineering*, vol. 9, no. 2, pp. 237–249, 2012.

[25] S. P. Boyd, L. E. Ghaoui, E. Feron, and V. Balakrishnan. Linear Matrix Inequalities in System and Control Theory. *Control Engineering Practice*, vol. 15, 1994.

[26] Y. Shtessel, C. Edwards, L. Fridman, and A. Levant. *Sliding mode control and observation.* Springer Science, 1st edition ed., 2014.

[27] M. Molaie, S. Jafari, J. C. Sprott, S. Golpayegani, and M. R. Hashemi. Simple Chaotic Flows with One Stable Equilibrium. *International Journal of Bifurcation and Chaos*, vol. 23, no. 11, p. 1350188, 2013.

Chapter 6

Generalised Projective Synchronisation between Lumped Parameter Modelled TLFMs

In this chapter, a generalised projective synchronisation (GPS) between a controlled master and multiple two-link flexible manipulators as slaves is developed. Here, two slave flexible manipulators are considered for illustration. The master and slave manipulators are nonidentical. However, all the slave manipulators are assumed to be identical to each other. The synchronisation is achieved using relative tracking errors between the master and slave manipulators. Complete and projective synchronisations are achieved for two different desired trajectories. These are exponentially varying and chaotic signals. An equivalent SMC is designed for the master manipulator to follow the desired trajectories. Two identical slave flexible manipulators are synchronised through some scaling factors to the master manipulator by designing a modified adaptive equivalent sliding mode control (MAE-SMC). The time varying sliding surface gain and time varying corrective control gain are designed using an adaptive SMC. The parameters of both the slave manipulators are considered unknown. Stability of the proposed technique is analysed using Lyapunov stability theory. It is shown that both the slave manipulators are synchronised properly with the master and thus, follow the desired trajectories. The link deflections are also suppressed within the desired value. A comparison of the proposed method (modified adaptive equivalent SMC with $tanh(s/\epsilon)$ $(MAE - SMC^2)$) with (i) a normal SMC with $tanh(s/\epsilon)$ (N-SMC), (ii) an adaptive equivalent SMC with $tanh(s/\epsilon)$ (AE-SMC) and (iii) a modified adaptive equivalent SMC with sigmoid $(s/|s| + \epsilon)$ $(MAE - SMC^1)$, where 's' is the sliding surface, is done.

The chapter is organised as the chapter objectives, the design of an equivalent sliding mode control and the design of a modified adaptive equivalent sliding mode control for synchronisation. These designed controllers and the lumped parameter model TLFM, derived in Chapter 3, are used to simulate the synchronisation scheme. The obtained simulation results are analysed and presented. Finally, the chapter summary is presented at the end.

6.1 Introduction

Robot manipulators are used in many industrial applications such as welding, manufacturing and assembly [1, 2]. Industrial robot manipulators are typically classified into two types: serial and parallel robot manipulators [2]. Usage of each type of manipulator depends on the purpose and the required specifications. The limitations and advantages of both types of manipulators are highlighted in [2]. But, a single robot alone may not be able to perform a specific complex and multitasking job. Hence, to increase the efficiency, it requires multiple or parallel robots to perform the given task. Different manipulators operate on a common function using the synchronisation phenomenon. The efficiency of the operation can be further enhanced by using flexible manipulators [3, 4]. Flexible manipulators have a number of advantages as compared to rigid manipulators [3, 5–7].

Various synchronisation strategies of robot manipulators are available in [8, 9]. Different types of synchronisation phenomenon are available in the literature for synchronisation between two-link rigid manipulators. They are motion synchronisation [1, 10], task space synchronisation [11–13], trajectory synchronisation [2, 14, 15] and force synchronisation [16, 17]. Identical robot manipulators are used both as the master and slave for the synchronous operation in all the above-mentioned papers. The synchronisation of the 3-R parallel robot for trajectory synchronisation is presented in [2]. Synchronisation of three robot manipulators is presented in [14]. Synchronisation for tracking control of 6-DOF hydraulic parallel manipulator is presented in [18]. Asymptotical synchronisation between two structurally different master and slave manipulators is presented in [19]. Motion synchronisation between two identical planar 3-DOF manipulators is presented in [20]. Trajectory tracking control of five-bar planar parallel manipulators is presented in [21]. Coordination motion control in the task space for the parallel manipulator is designed in [12]. Task space synchronisation control for two robot manipulators is presented in [13]. Motion synchronisation of two robot manipulators attached to one flexible beam is presented in [1]. Synchronisation of force for three robots is presented in [17]. Motion/force control for three two-link robot manipulators is presented in [22].

The synchronisation of the robot manipulators is applied in many fields like a bilateral teleoperation system [23]. In the case of a bilateral teleoperation system, a master manipulator is operated by humans and provide motion commands to the slave manipulator performing the actual task at a far place [24]. The stability and transparency (communication between master and slave) are considered as the main control objectives in a teleportation system. Categorisation of the papers for the synchronisation are given in Table 6.1. Many control techniques have been developed in the last decade to achieve the above-mentioned control objectives such as the PD type control technique

Table 6.1: Categorisation of the synchronisation papers based on the type of manipulators.

Sl. No.	Type of manipulators- Rigid/flexible	Synchronisation problem	References of paper
1.	Rigid manipulator	Motion synchronisation	[1, 22, 24]
2.	Rigid manipulator	Force synchronisation	[17, 25]
3.	Rigid manipulator	Task space/ trajectory synchronisation	[2, 14, 26]
4.	Rigid manipulator	Position synchronisation	[27]
5.	Rigid manipulator	Synchronization of networked robotic systems	[28]
6.	Rigid manipulator	Constrained distributed cooperative synchronisation	[29]
7.	Rigid manipulator	Temporal synchronisation	[30]
8.	Flexible joint	Trajectory synchronisation	[31]
9.	Two-link flexible – LPM modelled	Projective synchronisation	

[32, 33], direct force feedback control and passivity based control techniques [34]. Some robust control techniques are also developed for the above-mentioned control objectives like adaptive control in [23, 35], SMC in [36], predictive control using LMI [37]. In the past few years, many variants of SMC and adaptive control techniques are reported like adaptive fuzzy logic control (FLC) in [38], terminal SMC (TSMC) in [39] and adaptive neural network based nonsingular fast terminal SMC (NFTSMC) in [27]. An extended state observer based NFTSMC is reported in [40] for the master slave teleportation system. An adaptive control technique can work well in the presence of bounded disturbances and structured uncertainties. But, an adaptive control has some disadvantages as discussed in [41]. A sliding mode control technique (SMC) is robust to uncertainties, disturbances and gives faster responses [42]. The major disadvantage of a SMC control technique is the chattering problem [42]. The disadvantages of SMC and adaptive control techniques are overcome by using an adaptive SMC as noted in [41].

Motivation and problem formulation:

It is observed from Table 6.1 that various synchronisations are conducted between the rigid manipulators but not between flexible manipulators. In this book, a Lyapunov based equivalent sliding mode control is used for the master and a modified adaptive equivalent sliding mode control technique is proposed for the synchronisation of the TLFMs. The stability proof of both the techniques is also given. The design of an equivalent sliding mode controller involves choosing tunable sliding mode parameters. An adaptation law is developed for the modified adaptive equivalent sliding mode controller.

6.2 Chapter objectives

To solve the problem, the following objectives are set:

- To design an equivalent sliding mode controller for the master manipulator for tracking the desired exponentially varying and chaotic signals.

- To design a modified adaptive equivalent sliding mode controller for the synchronisation of the controlled master and the two slaves which are non-identical to the master manipulator.
- To design the above controllers in the presence of variable payloads and disturbances.
- To compare the performances of the designed controllers with the other three controllers.

6.3 Design of an equivalent SMC for the master manipulator

The structure of the controlled master-slave projective synchronisation strategy of TLFMs is shown in Fig. 6.1. An equivalent SMC technique is developed to track the desired signal by the master manipulator. Consider a master (TLFM) dynamics as

$$M_m \ddot{q}_m + D_m \dot{q}_m + K_m q_m = Q_m \tau_m \tag{6.1}$$

where $q_m = [\theta_{1m}, \theta_{2m}, \alpha_{1m}, \alpha_{2m}]^T$.

Let $q_d \epsilon R^4, q_d = [\theta_{1d}, \theta_{2d}, \alpha_{1d}, \alpha_{2d}]^T$ be the twice differentiable desired trajectories and allowable deflections for the master manipulator (6.1). So, the angular tracking and link deflection error is defined as

$$e_{im} = q_m - q_d \tag{6.2}$$

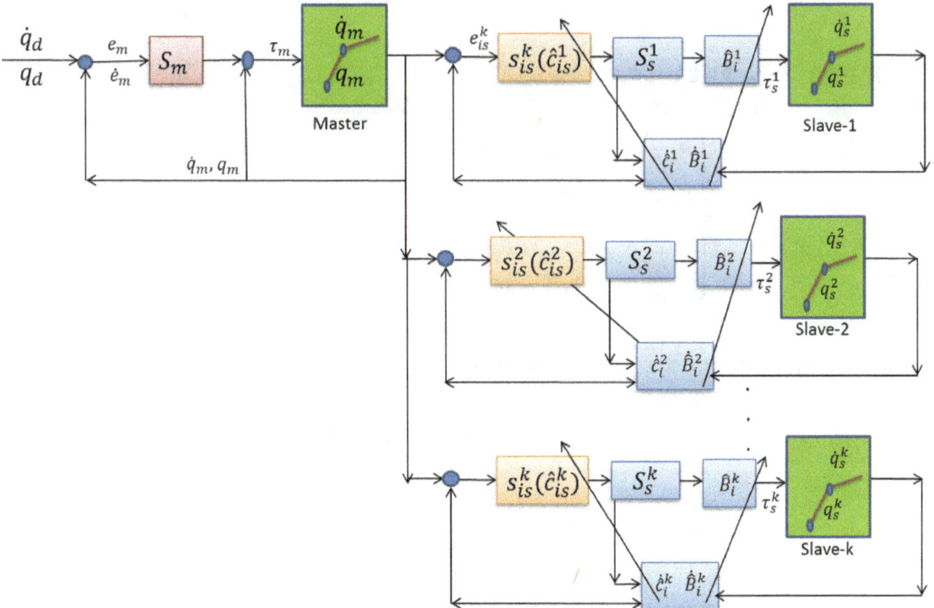

Figure 6.1: The master-slave manipulators synchronisation strategy.

Now, the objective is to design an equivalent SMC (S_m) for the master system (6.1) to track the desired trajectory θ_{id} and required link deflection α_{id}. Hence, $\|e_{im}\| \to 0$ as $t \to \infty$. Consider the dynamics of k slave manipulators (TLFM) described as :

$$M_s^k \ddot{q}_s^k + D_s^k \dot{q}_s^k + K_s^k q_s^k = Q_s^k u_s^k \tag{6.3}$$

where $k = 1, 2, ..., n$ and $q_s^k = [\theta_{1s}^k, \theta_{2s}^k, \alpha_{1s}^k, \alpha_{2s}^k]^T$ is the state of the kth slave manipulator. It is required to design a controller so that k slave manipulators completely or projectively follow the controlled master dynamics. Here, q_m is used as the reference trajectory for k slave manipulators. Thus, the projective synchronisation errors between the master (6.1) and slaves (6.3) are defined as:

$$e_{is}^k = q_{is}^k - \beta^k q_{im} \tag{6.4}$$

where β^k is the scaling factor used for projecting the slave trajectory to the master trajectory. Now, the objective is to design a modified adaptive equivalent sliding mode control (MAE-SMC) u_s^k for the k slave manipulators to track the desired master trajectory. Hence, the required goal is to achieve $\|e_{is}^k\| \to 0$ as $t \to \infty$. Thus, the expected synchronisation is required to guide the slaves only and the output of the master (q_m) is used for all the slave manipulators.

We can write from (6.1) as

$$\ddot{q}_m = M_m^{-1}(-D_m \dot{q}_m - K_m q_m + Q_m \tau_m) \tag{6.5}$$

Solving equation (6.5), we get

$$\begin{cases} \ddot{\theta}_{1m} = -p_1 \dot{\theta}_{1m} + p_2 \alpha_{1m} + p_3 \tau_m + d_{1m} \\ \ddot{\theta}_{2m} = -p_4 \dot{\theta}_{2m} + p_5 \alpha_{2m} + p_6 \tau_m + d_{2m} \\ \ddot{\alpha}_{1m} = p_1 \dot{\theta}_{1m} - p_7 \alpha_{1m} - p_3 \tau_m + d_{3m} \\ \ddot{\alpha}_{2m} = p_4 \dot{\theta}_{2m} - p_8 \alpha_{2m} - p_6 \tau_m + d_{4m} \end{cases} \tag{6.6}$$

where $p_{i=1,...,8}$ are the parameters of the system as given in (3.19) and values are determined using the parameters of the master TLFM given in Table 6.2. $d_{lm} = 0.2\sin(\dot{q}_{lm})\sin(q_{lm}), l = 1, 2, 3, 4$ are the disturbances associated with the plant dynamics (6.6). The disturbances are bounded as

$$\|d_{lm}\| < \|\dot{q}_{lm}\|\|q_{lm}\| \tag{6.7}$$

Let, θ_{id} be the twice differentiable desired trajectory and α be the desired deflection for the master system (6.6). Thus, the tracking error and deflection error are defined as:

$$\begin{cases} e_{im} = \theta_{im} - \theta_{id} \\ e_{jm} = \alpha_{im} - \alpha_{id} \end{cases} \tag{6.8}$$

where $i = 1, 2$ and $j = 3, 4$.

The expression of q_d is given in the results and discussion section. Here, the task of the controller is to operate on the system dynamics (6.5) so that the master system tracks the desired trajectory and satisfies the condition in (6.9).

$$\lim_{t \to \infty} \|e_{lm}(t)\| = 0, l = i, j \tag{6.9}$$

First, individual sliding surfaces are designed for all error dynamics, then an equivalent sliding surface is designed in terms of the individual surface to track the desired trajectory and to bring the dynamics on the surface.

$$s_{lm} = \dot{e}_{lm} + c_{lm} e_{lm} \tag{6.10}$$

$$S_m = s_{1m} + a_{1m} s_{2m} + a_{2m} s_{3m} + a_{3m} s_{4m} \tag{6.11}$$

where $a_{lm}, c_{lm} (l = 1, 2, 3, 4)$ are positive gains of the sliding surfaces whose values are defined by the user. The necessary and sufficient conditions for the existence of (6.11) is to operate the system on the sliding surface when it satisfies $S_m = \dot{S}_m = 0$. When the system operates on the sliding surface, $s_{lm} = 0$, the equivalent sliding mode dynamics for (6.10) are defined as

$$\ddot{e}_{lm} = -c_{lm} \dot{e}_{lm} \tag{6.12}$$

The stability of (6.12) can be ensured by considering a Lyapunov function candidate whose time derivative is given as $\dot{V}_m(\dot{e}_{lm}) = -c_{lm} \dot{e}_{lm}^2$. Hence, the trajectory of system dynamics (6.6) converges on (6.10). Now, the control input (τ_m) can be designed based on (6.11) so that the system tracks the desired trajectory as well as maintains on the sliding surface (6.11).

Theorem 6.1. *The system dynamics (6.6) reaches the sliding surface* $S_m = 0$ *in finite time* t_{rm} *and tracks the desired trajectory in presence of bounded disturbance (6.7) satisfying the desired condition (6.9) in approximated time along with the sliding surface (6.10), (6.11) by the control law given in (6.13).*

$$\begin{aligned} \tau_m = (-\frac{1}{T_m})\{&(A_1 + c_{1m}\dot{e}_{1m}) + a_{2m}(A_2 + c_{2m}\dot{e}_{2m}) + a_{3m}(A_3 + c_{3m}\dot{e}_{3m}) \\ &+ a_{4m}(A_4 + c_{4m}\dot{e}_{4m}) - \ddot{q}_d(a_{1m} + 1)\} - \phi_m tanh(\frac{S_m}{\epsilon}) \end{aligned} \tag{6.13}$$

where $T_m = p_3 + p_6 a_{1m} - p_3 a_{2m} - p_6 a_{3m}$, A_1, A_2, A_3, A_4 are the elements of (6.6), $tanh(S_m)$ is the tangent hyperbolic function in terms of S_m and ϕ_m is a user defined positive gain of the control input.

Proof. Consider a Lyapunov function candidate as $V_m(S_m) = 0.5\ S_m^2$ whose time derivative is given in (6.14).

$$\dot{V}_m(S_m) = S_m \dot{S}_m \tag{6.14}$$

Taking the time derivative of (6.11) and using the required control input (6.13) along with (6.6), (6.14) can be written as

$$\dot{V}_m(S_m) \le -\phi_m \frac{|S_m|}{\epsilon_m} \le -\phi_m |S_m| \tag{6.15}$$

Since $\phi_m > 0$ and $0 < \epsilon_m < 1$, (6.15) is a negative definite function according to the Lyapunov stability theory, asymptotic convergence of e_{lm} is guaranteed by (6.13) and converges to the sliding surface within the approximated time t_{rm} given in (6.16).

$$t_{rm} \le \frac{\mid S_m(0) \mid}{\phi_m} \tag{6.16}$$

For calculating the above equation, the individual sliding surfaces of the master manipulator for tracking the desired trajectory and bringing the dynamics onto the surface is designed as

$$s_{lm} = \dot{e}_{lm} + c_{lm} e_{lm} \tag{6.17}$$

where $l = 1, 2, 3, 4$.

The equivalent sliding surface is defined as

$$S_m = s_{1m} + a_{1m} s_{2m} + a_{2m} s_{3m} + a_{3m} s_{4m} \tag{6.18}$$

$$\dot{S}_m = \dot{s}_{1m} + a_{1m} \dot{s}_{2m} + a_{2m} \dot{s}_{3m} + a_{3m} \dot{s}_{4m} \tag{6.19}$$

where a_{lm}, c_{lm} are the positive gains of the sliding surface whose values are user defined.

For the S_m dynamics, a Lyapunov function candidate is introduced by taking the form

$$V_m = \frac{1}{2} S_m 2 \tag{6.20}$$

In order to provide the asymptotic stability of (6.19) about the equilibrium point $S_m = 0$, the following conditions must be satisfied:

- (a) $\dot{V}_m < 0$ for $S_m \neq 0$
- (b) $\lim_{\|S_m\| \to 0} V_m = \infty$

Condition (b) is obviously satisfied by V_m in (6.20). In order to achieve finite-time convergence (global finite-time stability) condition (b) can be modified to be

$$\dot{V}_m \leq -\phi_m V_m^{1/2}, \phi_m > 0 \tag{6.21}$$

Separating variables and integrating inequality (6.21) over the time interval $0 \leq \tau \leq t$, we obtain

$$\frac{dV_m}{V_m^{1/2}} \leq -\phi_m dt$$

$$\int_0^t \frac{dV_m}{V_m^{1/2}} \leq -\phi_m \int_0^t dt$$

$$V_m^{1/2}(t) - V_m^{1/2}(0) \leq -\frac{1}{2}\phi_m t$$

$$V_m^{1/2}(t) \leq -\frac{1}{2}\phi_m t + V_m^{1/2}(0) \tag{6.22}$$

Consequently, $V_m(t)$ reaches zero in finite time t_{rm} that is bounded by

$$-\frac{1}{2}\phi_m t_{rm} + V_m^{1/2}(0) \geq 0$$

$$t_{rm} \leq \frac{2V_m^{1/2}(0)}{\phi_m} \tag{6.23}$$

From (6.20),

$$V_m^{1/2} = \frac{|S_m|}{\sqrt{2}}$$

So,

$$V_m^{1/2}(0) = \frac{|S_m(0)|}{\sqrt{2}} \tag{6.24}$$

Putting (6.24) in (6.23),

$$t_{rm} \leq \sqrt{2}\frac{|S_m(0)|}{\phi_m} \tag{6.25}$$

Thus, error dynamics (6.8) satisfy the condition (6.9) and converges to the sliding surface $S_m = \dot{S}_m = 0$ in approximated time t_{rm}. Hence, the master manipulator tracks the desired trajectory. □

6.4 Design of a modified adaptive equivalent SMC for synchronisation between the controlled master manipulator and slave manipulators

Synchronisation of one controlled master and two slave TLFMs is presented in this section. Here, the controlled master is considered nonidentical to the slave manipulators, i.e., they have different sets of parameters. A modified adaptive equivalent SMC is designed for the synchronisation of manipulators. The objective for designing the proposed controller is to track the controlled master manipulator, (6.5) or indirectly the desired trajectories q_d. The term $d_{is}^k = 0.2\sin(\pi q_{is}^k)\sin(\pi \dot{q}_{is}^k)$ are the disturbances added to the dynamics (6.3) which are bounded as $\|d_{is}^k\| < \|q_{is}^k\| \|\dot{q}_{is}^k\|$.

6.4.1 Generalised projective synchronisation between nonidentical controlled master and k slave manipulators

Projective synchronisation means that the slave trajectories synchronise a scaling factor β times the master trajectories, i.e., the trajectories of the master and slave become proportional [43]. Considering q_m is the desired trajectory for all the k slave manipulators, generalised projective synchronisation (GPS) errors between the controlled master (6.5) and slaves (6.3) are defined as

$$e_{is}^k = q_{is}^k - \beta^k q_{im} \tag{6.26}$$

where β^k (> 0) is a user defined constant scaling factor used for GPS. The synchronisation of the master and k^{th} slave manipulator is achieved when (6.27) is satisfied.

$$\lim_{t \to \infty} \| e_{is}^k(t) \| = 0 \tag{6.27}$$

Now, the aim is to design a suitable controller such that the required condition in (6.27) is satisfied and each slave manipulator follows the master manipulator.

Modified adaptive equivalent SMC and stability analysis: First a sliding surface is designed for the individual error dynamics (6.26) followed by a design for a control law to maintain the required condition (6.27). The individual sliding surface is defined in (6.28).

$$s_{is}^k = \dot{e}_{is}^k + c_{is}^k e_{is}^k \tag{6.28}$$

Next an equivalent sliding surface in terms of the individual sliding surfaces is designed (6.28), for the whole system dynamics described as:

$$S_s^k = s_{1s}^k + a_{1s}^k s_{2s}^k + a_{2s}^k s_{3s}^k + a_{3s}^k s_{4s}^k \tag{6.29}$$

where c_{is}^k, $a_{is}^k (i = 1, 2, 3, 4) > 0$ are user defined sliding surface gains. Here, c_{is}^k is considered unknown. Now, the required control input u_s^k can be designed based on (6.29) so that each slave manipulator tracks the master manipulator and maintains on the sliding surface (6.29). The u_s^k is designed based on Theorem 6.2. Here, according to this control law, the parameters of the slave manipulators, gain of the sliding surfaces c_{is}^k, gain of the corrective control law ($\hat{\phi}$) are considered unknown. The value of the above unknown parameters is estimated. The parameters of the slave manipulators, the individual sliding surfaces gains and corrective control law gains are adjusted continuously during operation. Thus, the adaptation scheme is considered as indirect adaptive control.

Theorem 6.2. *Trajectories of the master and slave manipulators reach in an approximated time t^k_{rs} and maintain on the $S^k_s = \dot{S}^k_s = 0$. The kth slave manipulators (6.3) follow the master manipulator (6.5) satisfying the required conditions (6.27) using sliding surfaces (6.28), (6.29) and synchronisation controller (6.30) with adaptation laws (6.31), (6.32).*

$$\begin{aligned} u^k_s = (-\frac{1}{T^k_s})(&-\hat{b}^k_1\dot{\theta}^k_{1s} + \hat{b}^k_2\alpha^k_{1s} + A_1(\alpha_{1m},\dot{\theta}_{1m}) + \hat{c}^k_{1s}\dot{e}^k_{1s} + a^k_{1s}(-\hat{b}^k_4\dot{\theta}^k_{2s} + \hat{b}^k_5\alpha^k_{2s} \\ &+ A_2(\alpha_{2m},\dot{\theta}_{2m}) + \hat{c}^k_{2s}\dot{e}^k_{2s}) + a^k_{2s}(\hat{b}^k_1\dot{\theta}^k_{1s} - \hat{b}^k_7\alpha^k_{1s} + A_3(\alpha_{1m},\dot{\theta}_{1m}) \\ &+ \hat{c}^k_{3s}\dot{e}^k_{3s}) + a^k_{3s}(\hat{b}^k_4\dot{\theta}^k_{2s} - \hat{b}^k_8\alpha^k_{2s} + A_4(\alpha_{2m},\dot{\theta}_{2m}) + \hat{c}^k_{4s}\dot{e}^k_{2s})) - \hat{\phi}^k_s tanh(\frac{S^k_s}{\epsilon}) \end{aligned} \tag{6.30}$$

where $T^k_s = (b^k_3 + a^k_1 b^k_6 - a^k_2 b^k_3 - a^k_3 b^k_6)$ $\hat{b}^k_i(i = 1, 2, 4, 5, 7, 8)$. A_1, A_2, A_3, A_4 are the elements of (6.6). Here, $\hat{b}^k_i$ are the estimates of the parameters b^k_i, $\hat{c}^k_{is}$ are the estimates of sliding surface gain c^k_{is} and $\widehat{\phi}^k_s$ is the time varying corrective control gain, which are updated using the following adaptation laws:

$$\begin{cases} \dot{\hat{b}}^k_1 = -\dot{\theta}^k_{1s}S^k_s(1 + a^k_{2s}) + k^k_1\tilde{b}^k_1 \\ \dot{\hat{b}}^k_2 = \alpha^k_{1s}S^k_s(1 + a^k_{2s}) + k^k_2\tilde{b}^k_2 \\ \dot{\hat{b}}^k_4 = \dot{\theta}^k_{2s}S^k_s(a^k_{1s} - a^k_{3s}) + k^k_3\tilde{b}^k_4 \\ \dot{\hat{b}}^k_5 = \alpha^k_{2s}a^k_{1s}S^k_s + k^k_4\tilde{b}^k_5 \\ \dot{\hat{b}}^k_7 = -\alpha^k_{1s}a^k_{2s}S^k_s + k^k_5\tilde{b}^k_5 \\ \dot{\hat{b}}^k_8 = -\alpha^k_{2s}a^k_{3s}S^k_s + k^k_6\tilde{b}^k_8 \\ \tilde{b}^k_1 = b^k_1 - \hat{b}^k_1 \\ \tilde{b}^k_2 = b^k_2 - \hat{b}^k_2 \\ \tilde{b}^k_4 = b^k_4 - \hat{b}^k_4 \\ \tilde{b}^k_5 = b^k_5 - \hat{b}^k_5 \\ \tilde{b}^k_7 = b^k_7 - \hat{b}^k_7 \\ \tilde{b}^k_8 = b^k_8 - \hat{b}^k_8 \end{cases} \tag{6.31}$$

$$\begin{cases} \dot{\hat{c}}^k_1 = \dot{e}^k_{1s}, \dot{\hat{c}}^k_2 = \dot{e}^k_{2s}a^k_{1s} \\ \dot{\hat{c}}^k_3 = \dot{e}^k_{3s}a^k_{2s}, \dot{\hat{c}}^k_4 = \dot{e}^k_{4s}a^k_{3s} \\ \dot{\hat{\phi}}^k_s = -S^k_s tanh\left(S^k_s\right) \end{cases} \tag{6.32}$$

where k^k_i (> 0) are constant gains.

Proof. Considering another Lyapunov function candidate, as in (6.33) for the stabilisation of master and slave error dynamics (6.26) and equivalent sliding surface (6.29)

$$V_s^k(S_s^k, \tilde{b}_i^k, \tilde{c}_{is}^k, \tilde{\phi}_s^k) = 0.5(S_s^k)^2 + 0.5 \sum_{i=1,-(3,6)}^{8} (\tilde{b}_i^k)^2 + 0.5 \sum_{i=1}^{4} \tilde{c}_{is}^k + 0.5(\tilde{\phi}_s^k)^2 \tag{6.33}$$

where

$$\tilde{b}_i^k = (b_i^k - \hat{b}_i^k), \tilde{c}_{is}^k = (c_{is}^k - \hat{c}_{is}^k), \widetilde{\phi}_s^k = \left(\phi_s^k - \widehat{\phi}_s^k\right) \tag{6.34}$$

are the estimation errors. The time derivative of (6.33) is given in (6.35)

$$\dot{V}_s^k \left(S_s^k,\ \tilde{b}_i^k,\ \tilde{c}_{is}^k, \widetilde{\phi}_s^k \right) = S_s^k \dot{S}_S^k + \sum_{i=1,-(3,6)}^{8} \tilde{b}_i^k \dot{\hat{b}}_i^k + \sum_{i=1}^{4} \tilde{c}_{is}^k \dot{\hat{c}}_{is}^k + \widetilde{\phi}_s^k \dot{\widehat{\phi}}_s^k \tag{6.35}$$

Taking the time derivative of (6.28) and using the required control input (6.30) with the adaptation laws (6.31) and (6.32), (6.35) can be written as

$$\dot{V}_s \left(S_s^k,\ \tilde{b}_i^k,\ \tilde{c}_{is}^k, \widetilde{\phi}_s^k \right) = -(\tilde{\phi}_s^k)^2 |(S_s^k)| - \sum_{i=1,-(3,6)}^{8} (\tilde{b}_i^k)^2 - \sum_{i=1}^{4} (\tilde{c}_{is}^k)^2 \tag{6.36}$$

Since, ϕ_s^k, c_{is}^k, $a_{is}^k > 0$, (6.36) is a negative definite function asymptotic convergence of e_{is}^k is guaranteed by (6.30) and the error dynamics converge to the sliding surface within an approximated time t_{rs}^k, as given in (6.37).

$$t_{rs}^k \leq \frac{\left|S_s^k(0)\right|}{\widehat{\phi}_s^k} \tag{6.37}$$

Hence, error dynamics (6.26) satisfy the condition of (6.27) and converge to the sliding surface $S_s^k = \dot{S}_s^k = 0$ in an approximated time t_{rs}^k. Thus, the slave manipulators follow the master manipulator or the desired trajectory and GPS is achieved successfully. □

6.5 Results and discussion

Simulation results are discussed in this section. Here, one master and two slave flexible manipulators are used for the generalised projective synchronisation. The master and slave manipulators are nonidentical. However, the two slave manipulators are same and have identical parameters. The parameters of the master are given in Table 6.2 and two slave manipulators are given in Table 3.1. The desired reference signals used for the master flexible manipulator are given in (6.38) and (6.39). Two desired signals are used; one is exponentially varying (6.38) and the other is the chaotic system signal as in (6.39).

$$\theta_{id} = 1.2 - \left(\frac{7}{5}\right) e^{-t} + \left(\frac{7}{20}\right) e^{-4t}, \ i = 1, 2 \tag{6.38}$$

Table 6.2: Parameters of the TLFM (master (6.5)) [44].

Parameter	Link-1	Link-2
Length of links	$L_1 = 0.3\,m$	$L_2 = 0.2\,m$
Mass of links	m_1=0.065 kg	m_2=0.070 kg
Resistance of armature	$R_{m1} = 2.6\,\Omega$	$R_{m2} = 2.6\,\Omega$
Equivalent M.I. at load	$J_{eq1} = 0.099\,kgm^2$	$J_{eq2} = 0.092\,kgm^2$
M.I.	$J_{L1} = 0.00195\,kgm^2$	$J_{L2} = 0.000933\,kgm^2$
Coefficient of viscous damping	$B_{eq1} = 1.99\,Nms/rad$	$B_{eq2} = 1.99\,Nms/rad$
Efficiency of gear box	$\eta_{g1} = 0.9$	$\eta_{g2} = 0.9$
Efficiency of motor	$\eta_{m1} = 0.69$	$\eta_{2m} = 0.69$
Back e.m.f. constants	$K_{m1} = 0.00767\,V/rad$	$K_{m2} = 0.00767\,V/rad$
Gear ratios	$K_{g1} = 70$	$K_{g2} = 70$
Motor torque constants	$K_{t1} = 0.119\,Nm/A$	$K_{t2} = 0.0234\,Nm/A$
Natural frequency	$f_c = 3.2Hz$	
Stiffness of link	$K_{s1} = w_n^2 J_{eq1}\,Nm/rad$	$K_{s2} = w_n^2 J_{eq2}\,Nm/rad$

Here, the desired trajectories for both the links are considered the same, i.e., $\theta_{1d} = \theta_{2d} = \theta_d$. The dynamics of the chaotic system used for generating the desired reference trajectory is given in (6.39). Here, the Genesio-Tesi chaotic [45] system is used.

$$\begin{cases} \dot{y}_1 = y_2 \\ \dot{y}_2 = y_3 \\ \dot{y}_3 = r_3 y_3 + r_2 y_2 + r_1 y_1 + y_1^2 \end{cases} \tag{6.39}$$

where y_1, y_2, y_3 are the state variables and $r_1 = -1$, $r_2 = -1.1$, $r_3 = -0.45$ are the parameters of the system (6.39) used for chaotic behavior. We consider the chaotic desired trajectory as $\theta_d = y_1$, $\dot{\theta}_d = y_2$, $\ddot{\theta}_d = y_3$. Dynamics of the master (6.5) and slave manipulators (6.3) are simulated using ode-45 solver with a step size of 10^{-3} in a MATLAB® simulation environment. The actual parameter values of the slave manipulators dynamics are: b_1= 75.6721, b_2= 3.2358, b_3= 4.3867, b_4= 1026.12, b_5= 387.6975, b_6= 59.4266, b_7= 274.2314 and b_8= 3883.41. The actual value of the constants used for sliding surfaces and controllers for both the slave manipulators is: $c_{1s}^1 = c_{1s}^2 = 10, c_{2s}^1 = c_{2s}^2 = 7, \phi_s^1 = \phi_s^2 = 5, c_{3s}^1 = c_{3s}^2 = 0.5, c_{4s}^1 = c_{4s}^2 = 6, a_{1s}^1 = a_{1s}^2 = 0.6, a_{2s}^1 = a_{2s}^2 = 0.4, a_{3s}^1 = a_{3s}^2 = 0.4$ and $\varepsilon = 1$. The initial conditions for simulating the different dynamical equations are given in Table 6.3. The responses of the desired signals in (6.38) and (6.39) used for the synchronisation are shown in Fig. 6.2.

6.5.1 Trajectory tracking of an exponentially varying signal by the master manipulator

The value of constant gains of sliding surfaces and controller gain for controlling the master manipulator are considered as: $c_1 = 7$, $c_2 = 2$, $\phi_m = 5$, $c_3 = 2$, $c_4 = 0.05$, $a_1 = 0.6$, $a_2 = 0.3$, $a_3 = 0.14$, $\alpha_{1d} = \alpha_{2d} = 0$. Time responses of the angular positions $(\theta_{1m}, \theta_{2m})$ of the master

Table 6.3: Initial conditions used for simulating different equations.

Initial conditions	Description
$q_m(0) = (0.5, 0.5, 0.001, 0.002, 0, 0, 0, 0)^T$	For master manipulator when q_d is (6.38)
$q_s^1(0) = (0.1, 0.1, 0.001, 0.002, 0, 0, 0, 0)^T$	For slave-1 manipulator when q_d is (6.38)
$q_s^2(0) = (0.01, 0.01, 0.001, 0.002, 0, 0, 0, 0)^T$	For slave-2 manipulator when q_d is (6.38)
$q_m(0) = (1.0, 1.0, 0.001, 0.002, 0, 0, 0, 0)^T$	For master manipulator when q_d is (6.39)
$q_s^1(0) = (0.5, 0.5, 0.001, 0.002, 0, 0, 0, 0)^T$	For slave-1 manipulator when q_d is (6.39)
$q_s^2(0) = (0.01, 0.01, 0.001, 0.002, 0, 0, 0, 0)^T$	For slave-2 manipulator when q_d is (6.39)
$\hat{b}_i^1(0), \hat{b}_i^2(0) = (0, 0, 0, 0, 0, 0)^T$	For estimating the parameters of the , i.e., of (6.31).
$\hat{c}_{is}^1(0), \hat{c}_{is}^2(0) = (15, 15, 5, 5)^T$	For estimating the gain of the sliding surface for the slaves, i.e., of (6.32).
$\hat{\phi}_s^1(0), \hat{\phi}_s^2(0) = 1, 1$	For estimating the gain of the corrective control law.

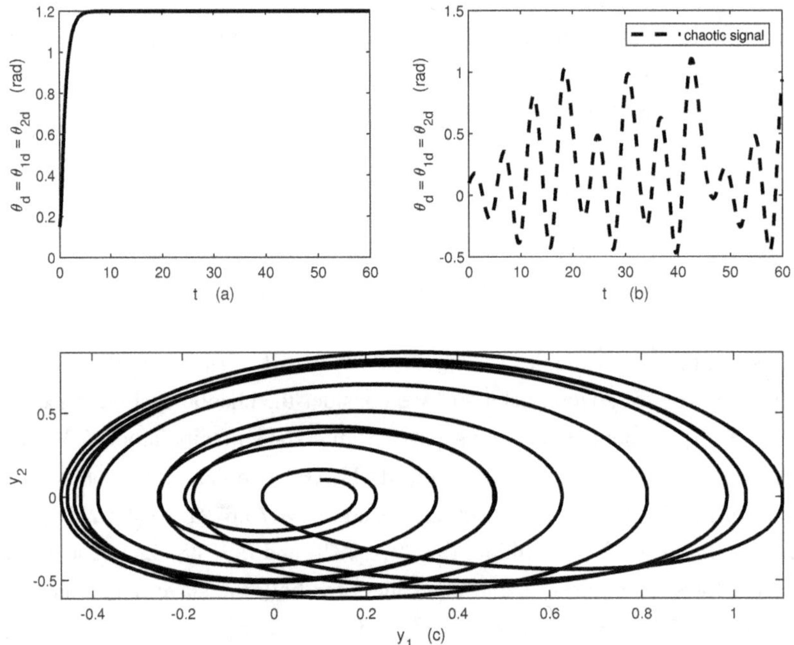

Figure 6.2: The desired trajectory tracking signals used: (a) exponentially varying as given in (6.38) and (b) low frequency chaotic signal in (6.39).

manipulator tracking the desired trajectories are shown in Fig. 6.3. The phase plane of the tracking error for the controlled master is given in Fig. 6.4. Time responses of the sliding surfaces and control inputs for controlling the master manipulator are shown in Fig. 6.5. Time responses of the deflection of the links for the master manipulator are given in Fig. 6.6. The responses of the individual sliding surfaces for the master manipulator are shown in Fig. 6.7. It is clear from Fig. 6.3 that both the links of the master manipulator are controlled properly and follow the desired trajectories (6.38) successfully. It is clear from Fig. 6.6 that the deflection of the links of master manipulator is

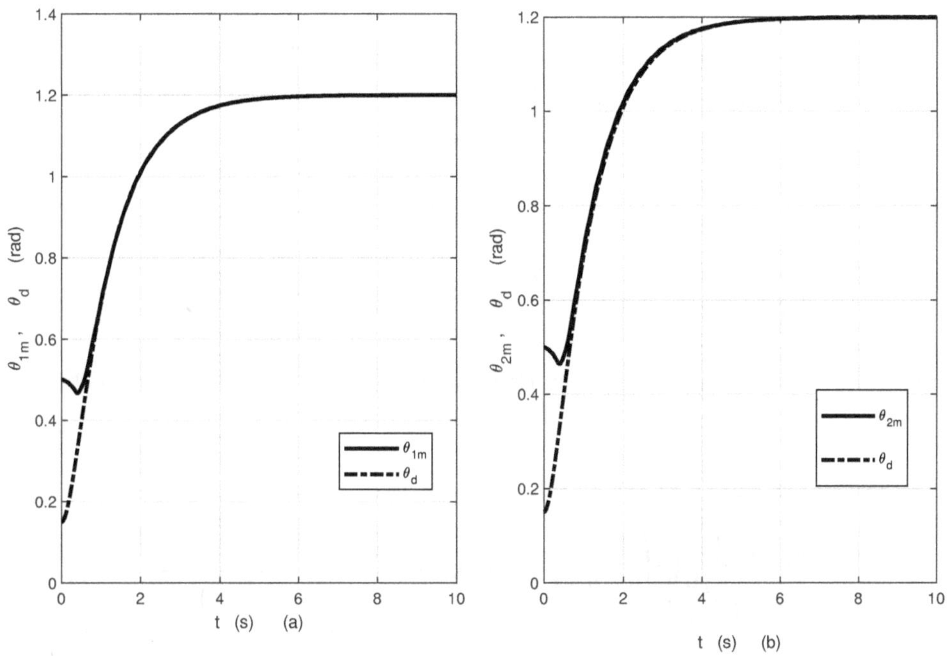

Figure 6.3: Tracking the angular position of the master manipulator when θ_d is (6.38): (a) link-1 and (b) link-2.

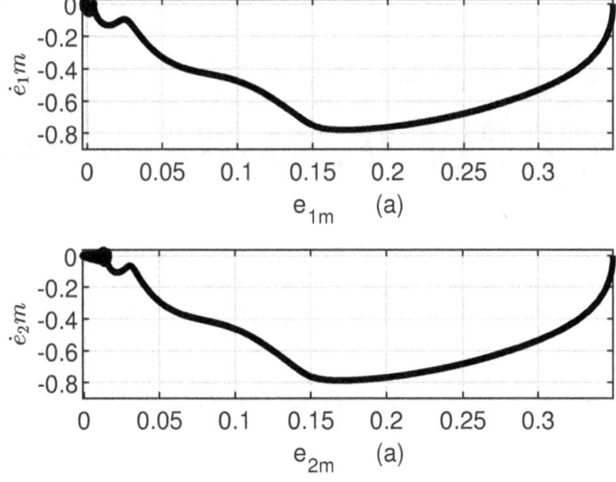

Figure 6.4: Phase plane of the tracking error for the controlled master when θ_d is (6.38): (a) on $e_{1m} - \dot{e}_{1m}$ (b) on $e_{2m} - \dot{e}_{2m}$.

suppressed properly within the desired value. It may be noted that the tip trajectory tracking of the master manipulator is successfully achieved along with the suppression of links deflection. The responses of the individual sliding surfaces of the master manipulator going to zero in Fig. 6.7 ensure that the equivalent sliding surface of the master has also gone to zero.

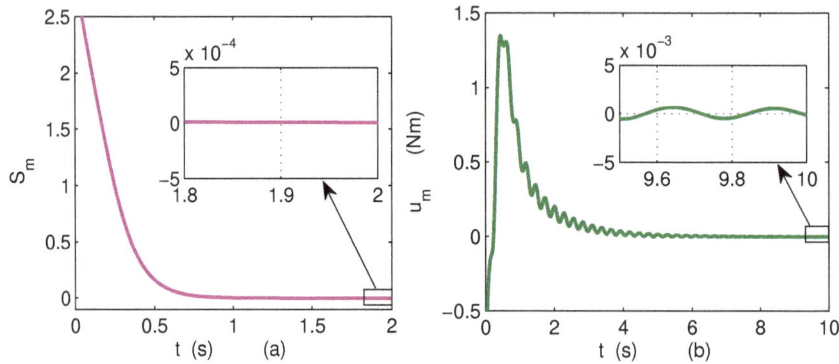

Figure 6.5: (a) Sliding surface (S_m) and (b) control input (u_m) when the master manipulator is set to track θ_d in (6.38).

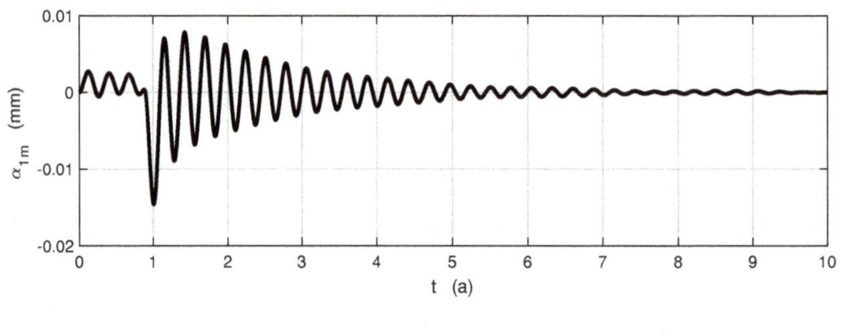

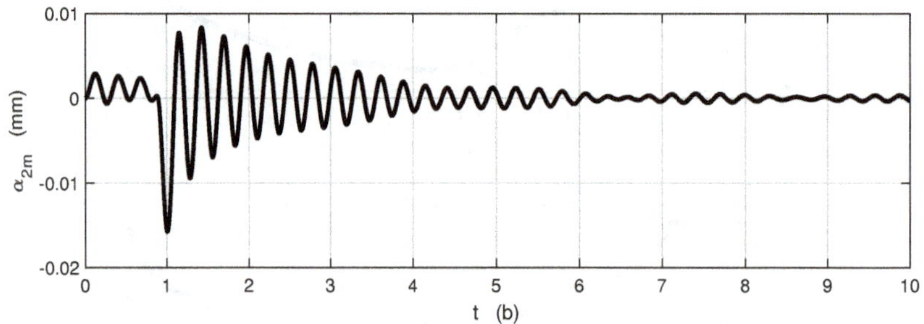

Figure 6.6: Link deflections of the master manipulator when θ_d is (6.38): (a) link-1 (α_{1m}) and (b) link-2 (α_{2m}).

6.5.2 Synchronisation of the controlled master and k slave manipulators

This section describes the synchronisation between the master manipulator (6.5) and two slave manipulators (6.3). Although two slave manipulators are identical to each other but the controlled master and slave manipulators are not. The objective is to track the master manipulator trajectories with proper suppression of links deflection when θ_d is (6.38). Here, two different scaling factors are used for synchronisation of slave manipulators with the master. The next two subsections describe the use of two scaling factors for GPS.

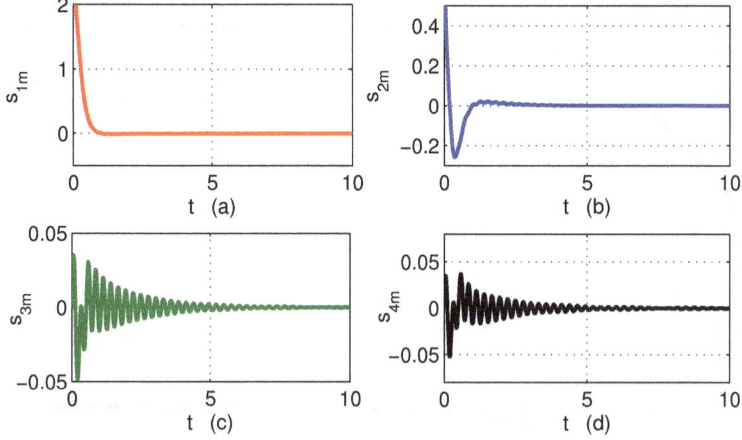

Figure 6.7: The individual sliding surfaces: (a) S_{1m}, (b) S_{2m}, (c) S_{3m} and (d) S_{4m}.

6.5.2.1 Scaling factor β^1=β^2=1 (complete synchronisation)

In this case, the synchronisation of the control master with two slave manipulators is considered complete because $\beta^1 = \beta^2 = 1$ for GPS. The time responses of the angular positions of the two slave manipulators tracking the master trajectories are shown in Fig. 6.8. The phase plane of the synchronisation errors between the master and slave-1 manipulators is shown in Fig. 6.9. The phase plane of the synchronisation errors between the master and slave-2 manipulators is shown in Fig. 6.10. It is clear from the Figs. 6.9 and 6.10 that the slave manipulators are tracking the desired controlled master properly and completely. The time responses of the equivalent sliding surfaces and control inputs for the two slave manipulators during synchronisation are shown in Fig. 6.11. The time responses of the links deflection of both the slave manipulators are shown in Fig. 6.12. Figure 6.12 indicates that the links deflection of both the slave manipulators are suppressed properly during synchronisation. Estimation of the parameters of both the slaves are shown in Fig. 6.13. It is apparent from the figure that the parameters of the slave manipulators converge to their actual value.

6.5.2.2 Scaling factor β^1=0.5, β^2=0.25

In this case, the slave-1 manipulator is synchronised (projected) to the controlled master with a scaling of $\beta^1 = 0.5$ and slave-2 manipulator is synchronised (projected) to the controlled master with a scaling of $\beta^2 = 0.25$. The angular positions of the slave manipulators tracking the desired master when θ_d is (6.38) are shown in Fig. 6.14. It is evident from Fig. 6.14 that the slave-1 manipulator is projected at a scaling of 0.5 to the master and slave-2 manipulators are projected at a scaling of 0.25 to the master. The phase plane of the synchronisation errors between the master and slave-1 manipulators is shown in Fig. 6.15. The phase plane of the synchronisation errors between the master and slave-2 manipulator are shown in Fig. 6.16. It is clear from Figs. 6.15 and 6.16 that the slave manipulators track the desired master properly. The time responses of the sliding surfaces and control inputs for the slave manipulators during synchronisation are shown in Fig. 6.17. The time

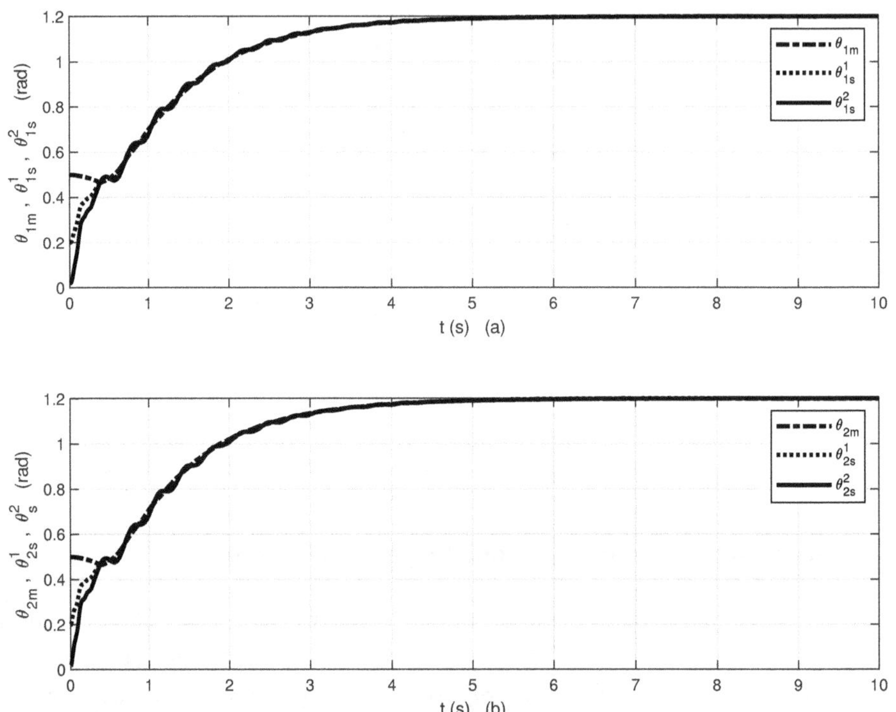

Figure 6.8: Synchronisation between the controlled master and two slave manipulators when θ_d is (6.38) with $\beta^1 = \beta^2 = 1$: (a) link-1 and (b) link-2.

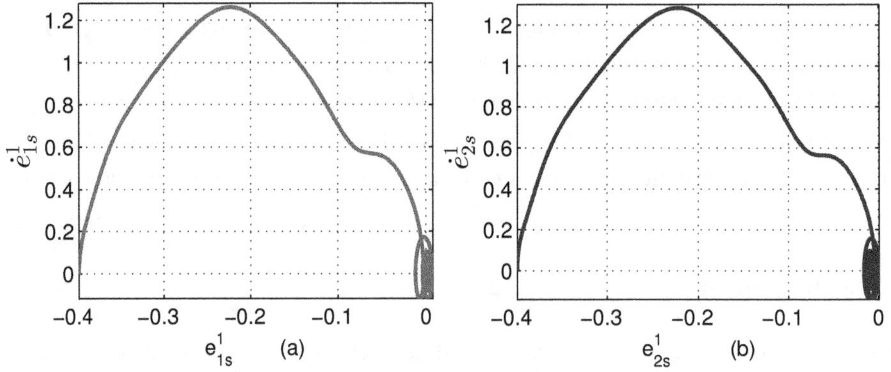

Figure 6.9: Phase plane of the synchronisation errors between the master and slave-1 when q_d is (6.38) with $\beta^1 = \beta^2 = 1$: (a) on $e^1_{1s} - \dot{e}^1_{1s}$ and (b) on $e^1_{2s} - \dot{e}^1_{2s}$.

responses of the links deflection of both the slave manipulators are shown in Fig. 6.18. Figure 6.18 depicts that the links deflection of both the slave manipulators are suppressed properly during synchronisation. Estimation of the parameters of both the slaves are shown in Fig. 6.19. It is observed that the parameters of the slave manipulators converge to their actual value.

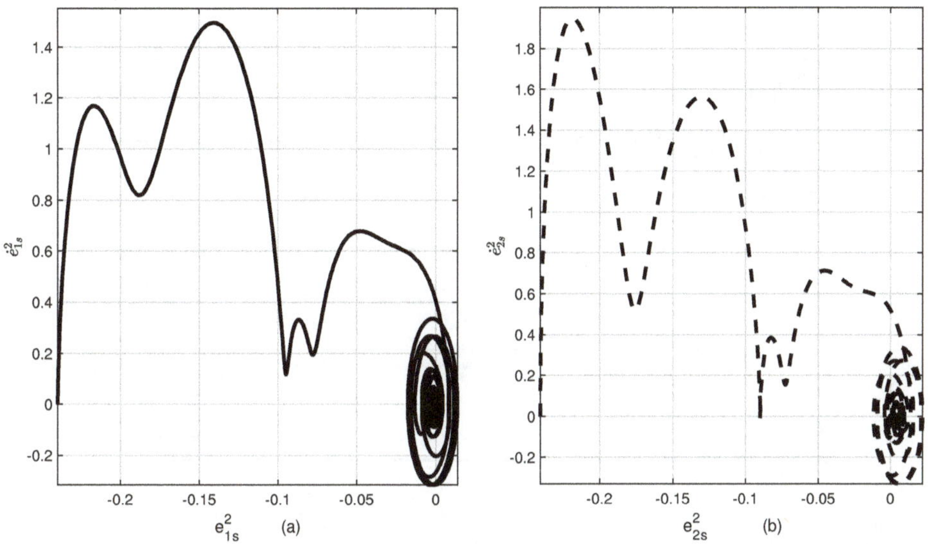

Figure 6.10: Phase plane of the synchronisation errors between the master and slave-2 when q_d is (6.38) with $\beta^1 = \beta^2 = 1$: (a) on $e^1_{1s} - \dot{e}^1_{1s}$ and (b) on $e^1_{2s} - \dot{e}^1_{2s}$.

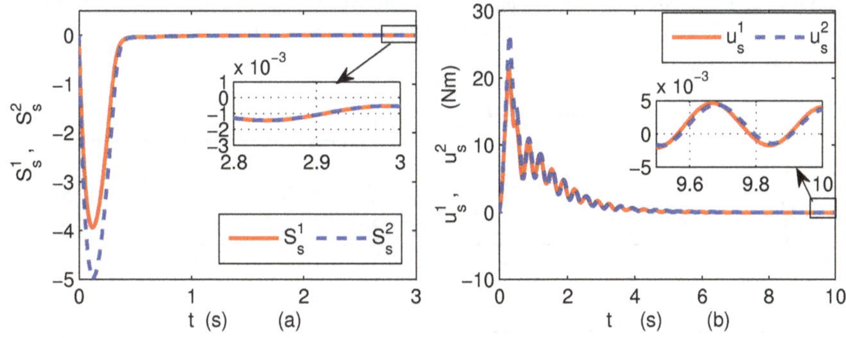

Figure 6.11: (a) Equivalent sliding surfaces (S^1_s, S^2_s) and (b) control inputs (u^1_s, u^2_s) for the synchronisation when θ_d is (6.38) with $\beta^1 = \beta^2 = 1$.

6.5.3 Trajectory tracking of a chaotic signal by the master manipulator

Here, the objective is to control the master manipulator to track the desired chaotic signal trajectory. But, extra challenges occur in the deflection of links due to the trajectory induced by the chaotic signal. A separate error is defined for suppression of links deflection to achieve the desired deflection, i.e., $\alpha = 0$. The angular positions of the master manipulator tracking the desired chaotic trajectory are shown in Fig. 6.20. The phase plane of the tracking error for the controlled master is given in Fig. 6.21. It is clear from Fig. 6.20 that the master manipulator is controlled properly and follows the desired chaotic trajectory (6.39) successfully. The time responses of the sliding surface and control input to track the master TLFM for tracking the chaotic trajectory are shown in Fig. 6.22. The time responses of the deflection of the links for the master manipulator are given in Fig. 6.23.

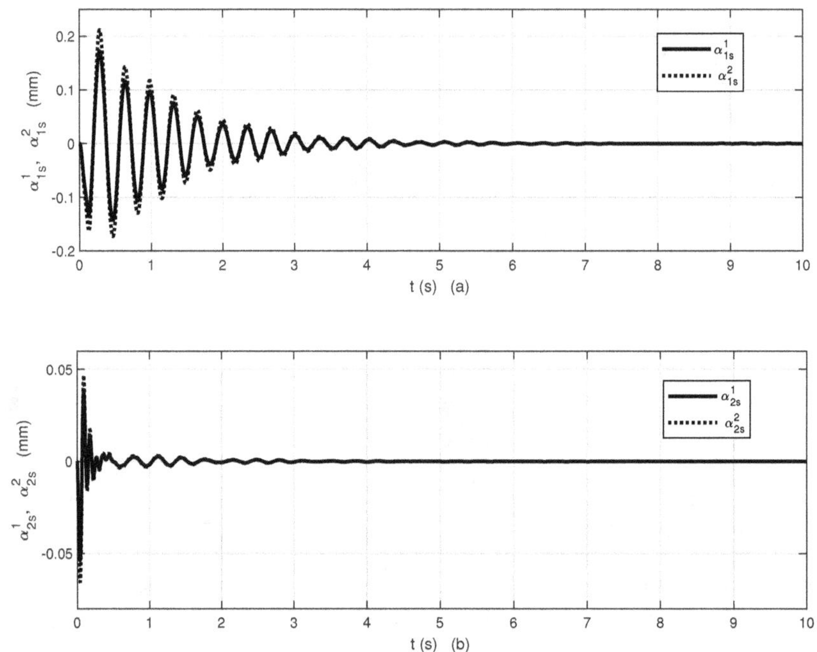

Figure 6.12: Link deflections of the slave manipulators during synchronisation when θ_d is (6.38) with $\beta^1 = \beta^2 = 1$: (a) link-1 and (b) link-2.

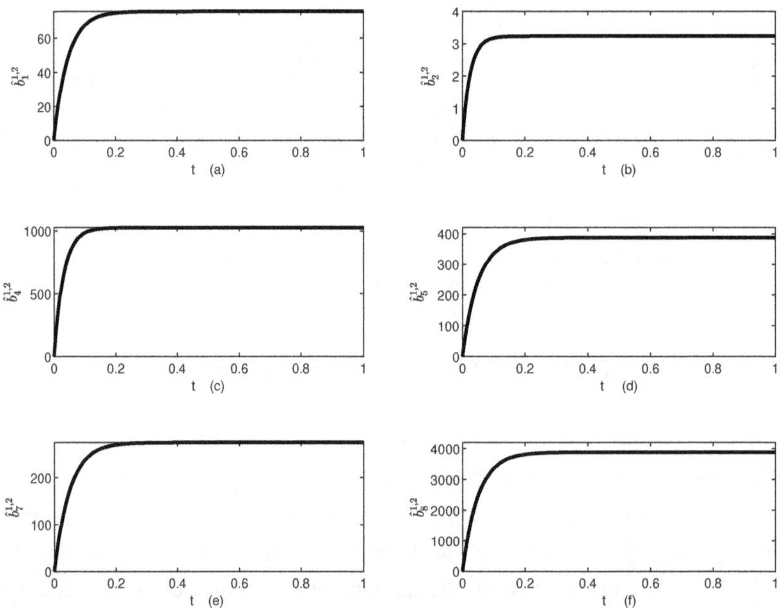

Figure 6.13: Time responses of the estimated parameters of the slave manipulators when θ_d is (6.38) with $\beta^1 = \beta^2 = 1$. $\hat{b}_i^{1,2}$ means $\hat{b}_i$ for slave-1 and slave-2.

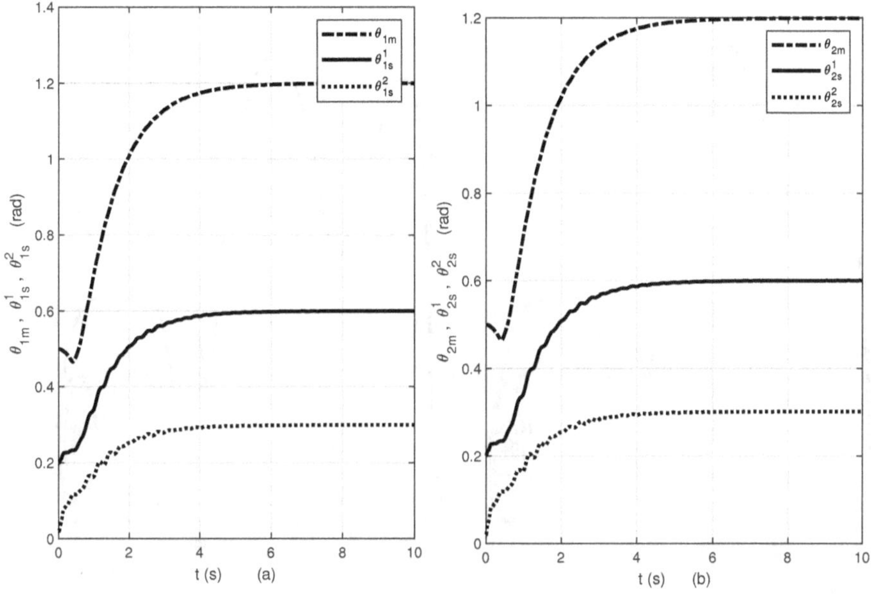

Figure 6.14: Projective synchronisation between the controlled master and slave manipulators with $\beta^1 = 0.5$, $\beta^2 = 0.25$ when θ_d is (6.38): (a) link-1 and (b) link-2.

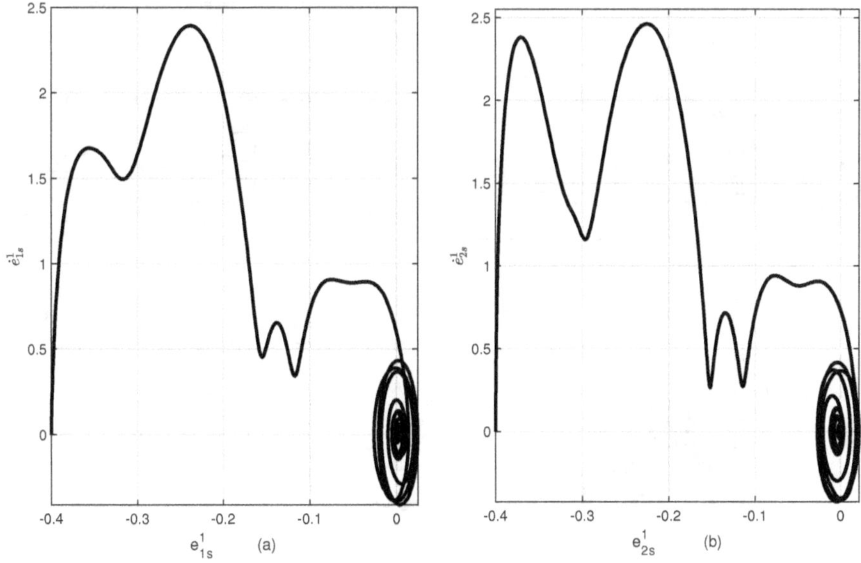

Figure 6.15: Phase plane of the synchronisation errors between the master and slave-1 manipulators when θ_d is (6.38) with $\beta^1 = 0.5, \beta^2 = 0.25$: (a) on $e^1_{1s} - \dot{e}^1_{1s}$ and (b) on, $e^1_{2s} - \dot{e}^1_{2s}$.

It is seen from Fig. 6.22 that the oscillation frequency in the transient phase for the control input gradually reduces. It is also seen from Fig. 6.23 that there is considerable deflection in the links during the transient phase, but gradually decays. It is noted from Fig. 6.23 that the links deflection of the master manipulator is suppressed properly and is within 0.005 mm.

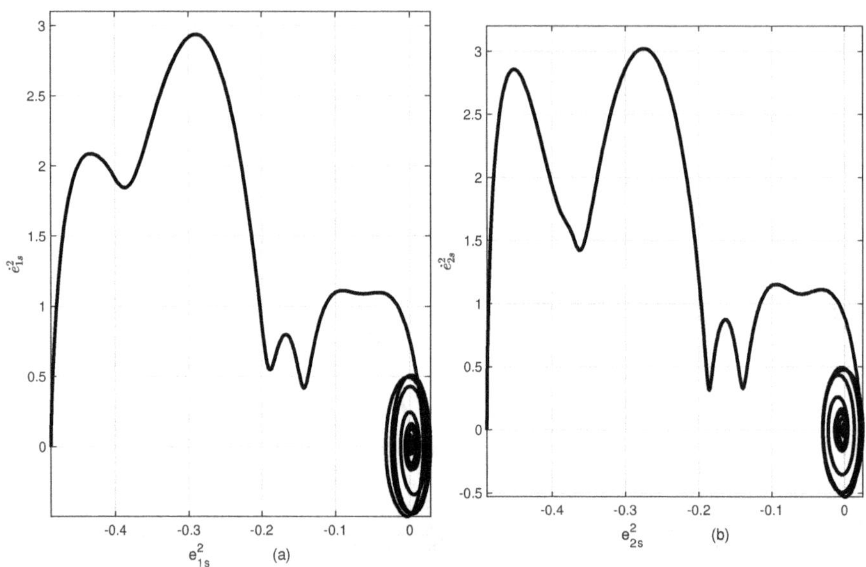

Figure 6.16: Phase plane of the synchronisation errors between the master and slave-2 manipulators when θ_d is (6.38) with $\beta^1 = 0.5$, $\beta^2 = 0.25$: (a) on $e^2_{1s} - \dot{e}^2_{1s}$ and (b) on $e^2_{2s} - \dot{e}^2_{2s}$.

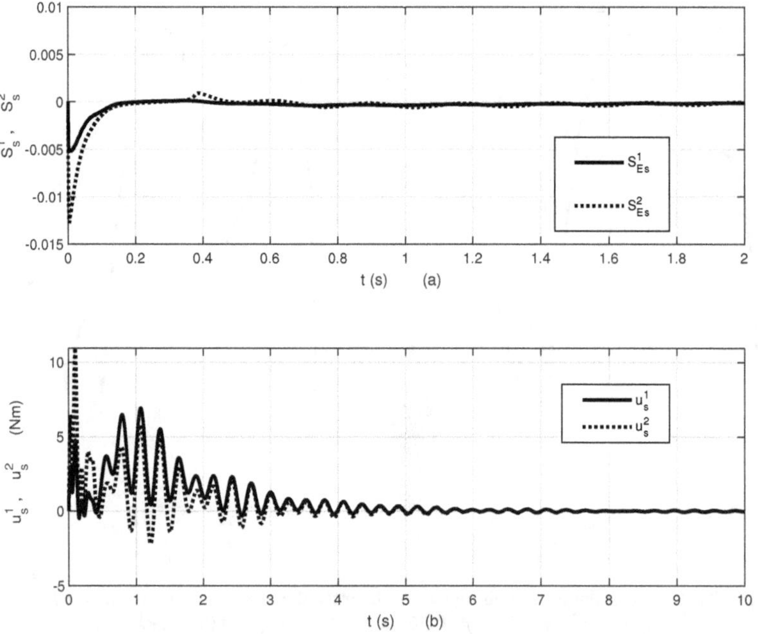

Figure 6.17: (a) Sliding surfaces (S^1_s, S^2_s) and (b) control inputs (u^1_s, u^2_s) for the synchronisation between master and slave manipulators with $\beta^1 = 0.5$, $\beta^2 = 0.25$.

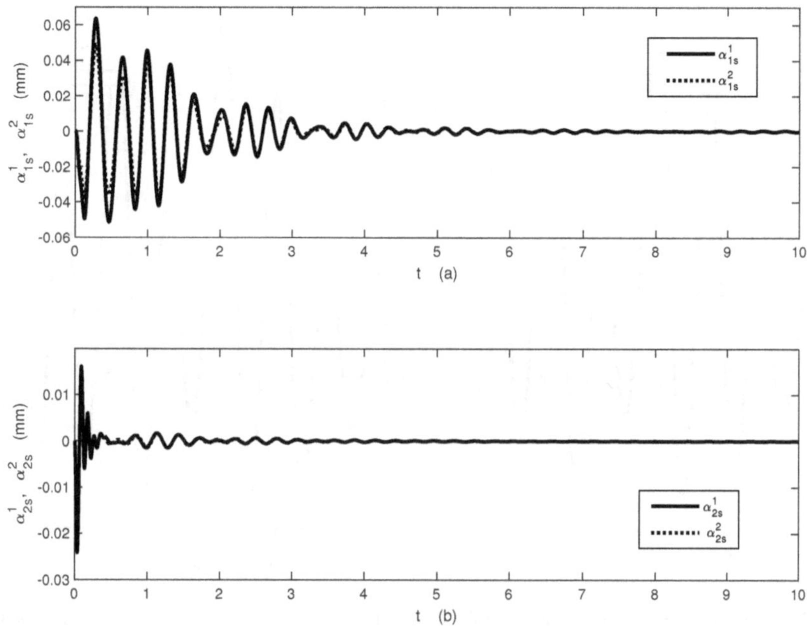

Figure 6.18: Link deflections of the slave manipulators during the synchronisation when θ_d is (6.38) with $\beta^1 = 0.5, \beta^2 = 0.25$: (a) link-1 and (b) link-2.

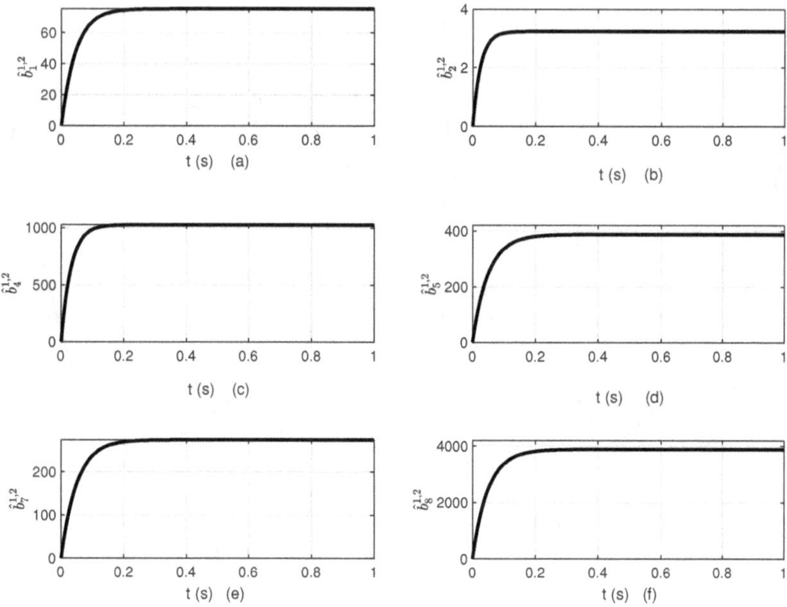

Figure 6.19: Time responses of the estimated parameters of the slave manipulators when θ_d is (6.38) with $\beta^1 = 0.5, \beta^2 = 0.25$.

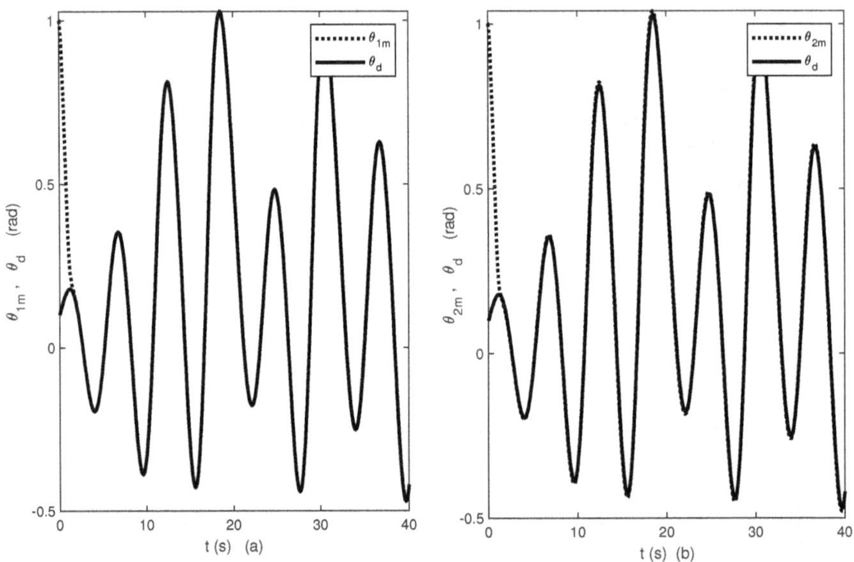

Figure 6.20: Tracking of the angular position of the master manipulator when θ_d is (6.39): (a) link-1 and (b) link-2.

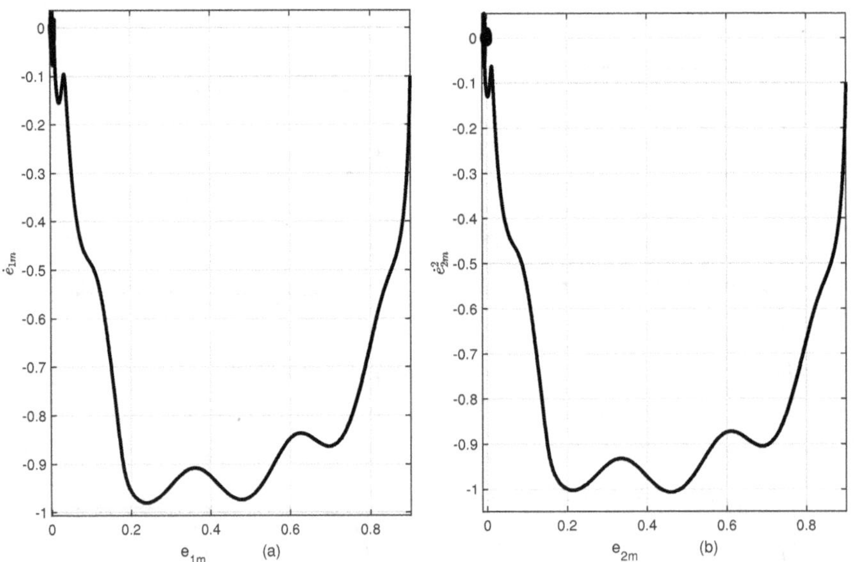

Figure 6.21: Phase plane of the tracking error for the controlled master when θ_d is (6.39) (a) on $e_{1m} - \dot{e}_{1m}$ and (b) on $e_{2m} - \dot{e}_{2m}$.

6.5.4 Synchronisation of master and k slave manipulators when desired trajectory θ_d is a chaotic signal and scaling factor β^1=0.5, β^2=0.25

Now, the objective is to synchronise the slave manipulators with the controlled master manipulator for the desired chaotic trajectory (6.39). The angular positions of the master and slave manipulators

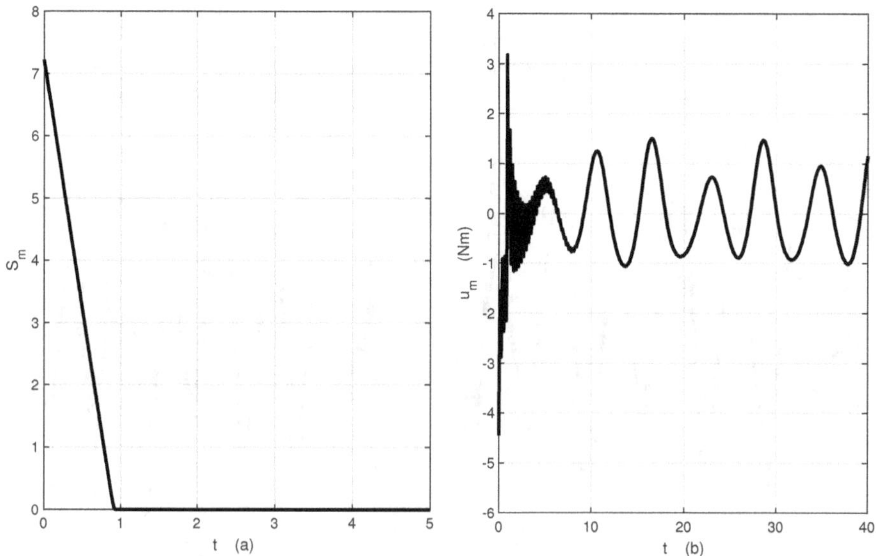

Figure 6.22: (a) Sliding surface (S_m) and (b) control input (u_m) when the master manipulator is set to track the desired chaotic signal in (6.39).

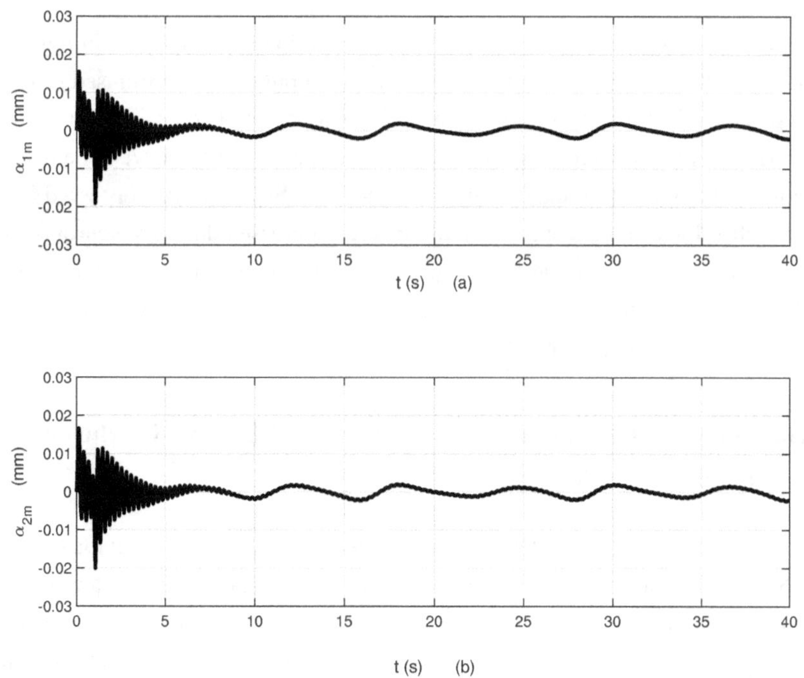

Figure 6.23: Link deflections of the master manipulator when θ_d is (6.39): (a) link-1 and (b) link-2.

with the desired chaotic trajectory are shown in Fig. 6.24. The phase plane of the synchronisation errors between the master and slave-1 manipulators is shown in Fig. 6.25. The phase plane of the

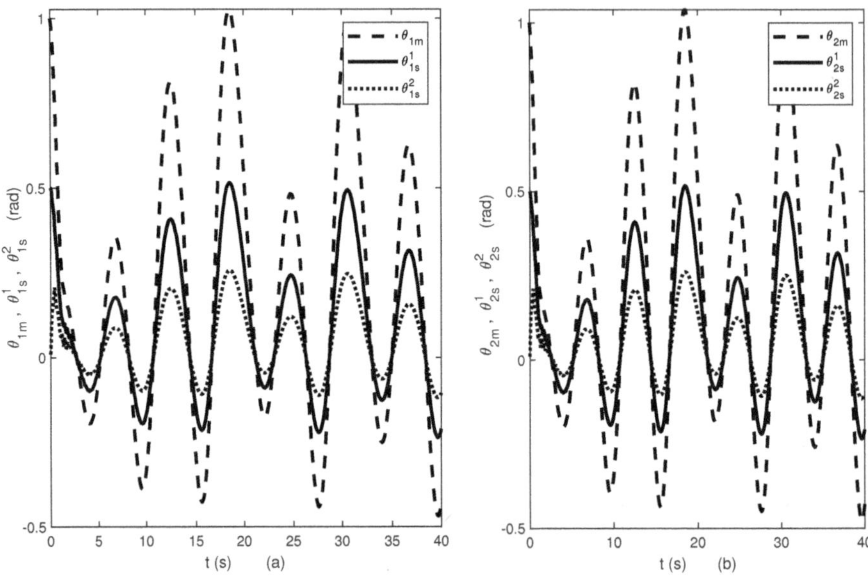

Figure 6.24: Projective synchronisation between the master and slave manipulators when θ_d is (6.39) with $\beta^1 = 0.5, \beta^2 = 0.25$: (a) link-1 (b) link-2.

synchronisation errors between the master and slave-2 manipulators is shown in Fig. 6.26. It is apparent from Fig. 6.25 and Fig. 6.26 that both the slave manipulators synchronised properly with the master manipulator. The time responses of the sliding surfaces and control inputs for the slave manipulators during chaotic trajectory synchronisation are shown in Fig. 6.27. Time responses of the links deflection of both the slave manipulators are shown in Fig. 6.28. It is observed from Fig. 6.28 that the links deflection of both slave manipulators are suppressed during synchronisation. Estimation of the parameters of the both the slave manipulators are shown in Fig. 6.29. It is observed from the Fig. 6.29 that the parameters of the slave manipulators converge to their actual value in the case of chaotic desired trajectory tracking also.

6.5.5 Synchronisation in the presence of payload of 0.45 kg for the chaotic desired trajectory

Before this, the synchronisation strategies are presented in the absence of payload in the slave manipulators. Now, a payload of $M_p = 0.45$ kg is added to the slave manipulators to check the robustness and usefulness of the synchronisation technique. In this case, the chaotic desired trajectory is used for the generalised projective synchronisation between the controlled master and slave manipulators. The GPS results in the presence of payload are presented now. The angular positions of the master and slave manipulators with chaotic desired trajectory in the presence of payload and scaling factors $\beta^1 = 0.5, \beta^2 = 0.25$ are shown in Fig. 6.30. The phase plane of the synchronisation errors between the master and slave-1 manipulators is shown in Fig. 6.31. The phase plane of the synchronisation errors between the master and slave-2 manipulator is shown in Fig. 6.32. It is

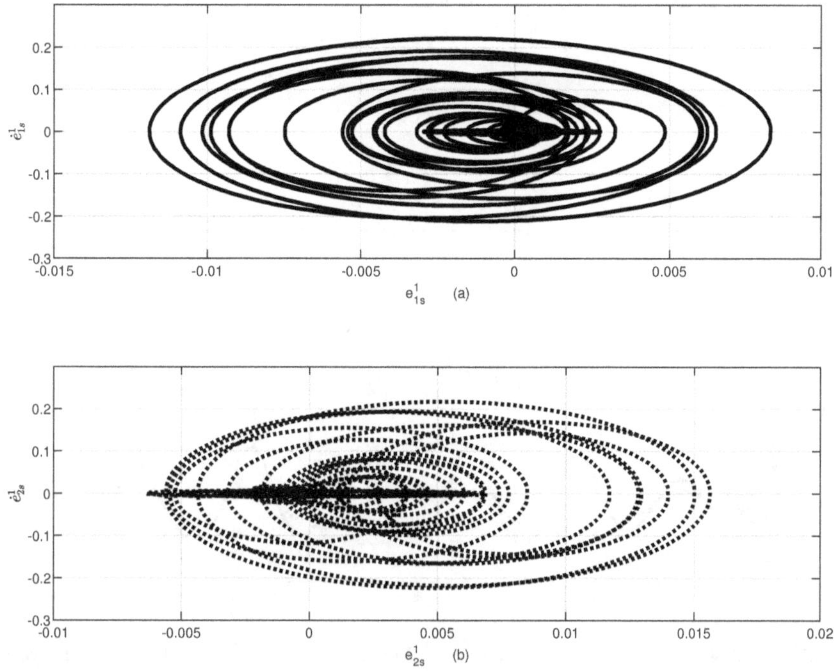

Figure 6.25: Phase plane of the synchronisation errors between the master and slave-1 when θ_d is (6.39) with $\beta^1 = 0.5, \beta^2 = 0.25$: (a) on $e_{1s}^1 - \dot{e}_{1s}^1$ and (b) on $e_{2s}^1 - \dot{e}_{2s}^1$.

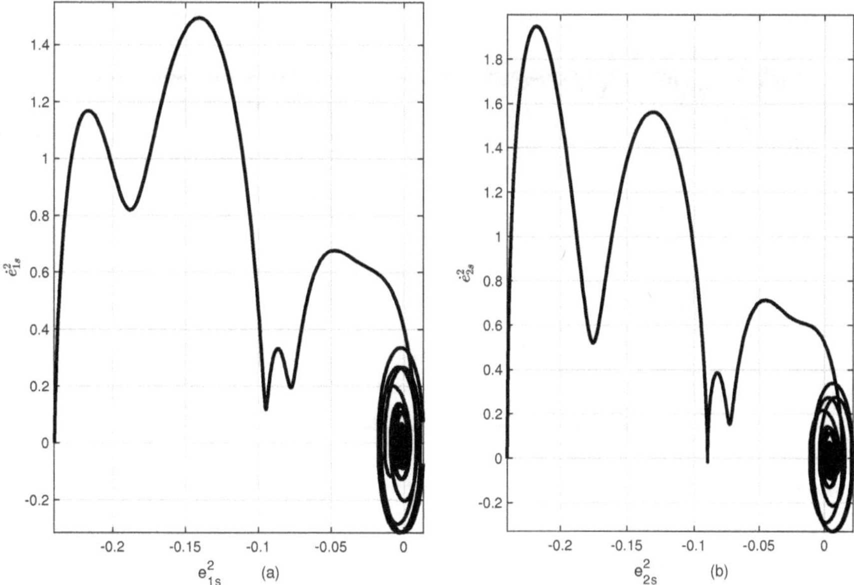

Figure 6.26: Phase plane of the synchronisation errors between the master and slave-2 when θ_d is (6.39) with $\beta^1 = 0.5, \beta^2 = 0.25$: (a) on $e_{1s}^2 - \dot{e}_{1s}^2$ and (b) on $e_{2s}^2 - \dot{e}_{2s}^2$.

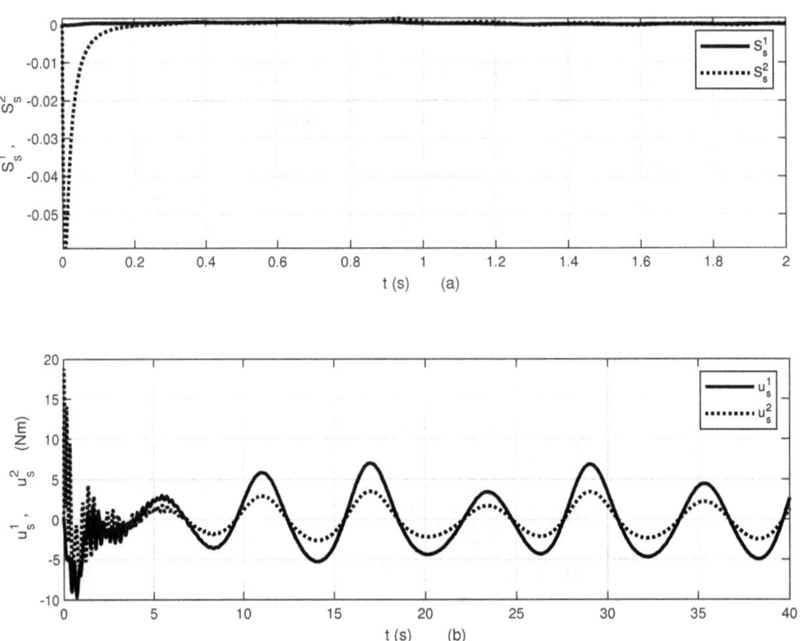

Figure 6.27: (a) Sliding surfaces (S_s^1, S_s^2) and (b) control inputs (u_s^1, u_s^2) for the synchronisation between the master and slave manipulators with $\beta^1 = 0.5$,$\beta^2 = 0.25$ when θ_d is (6.39).

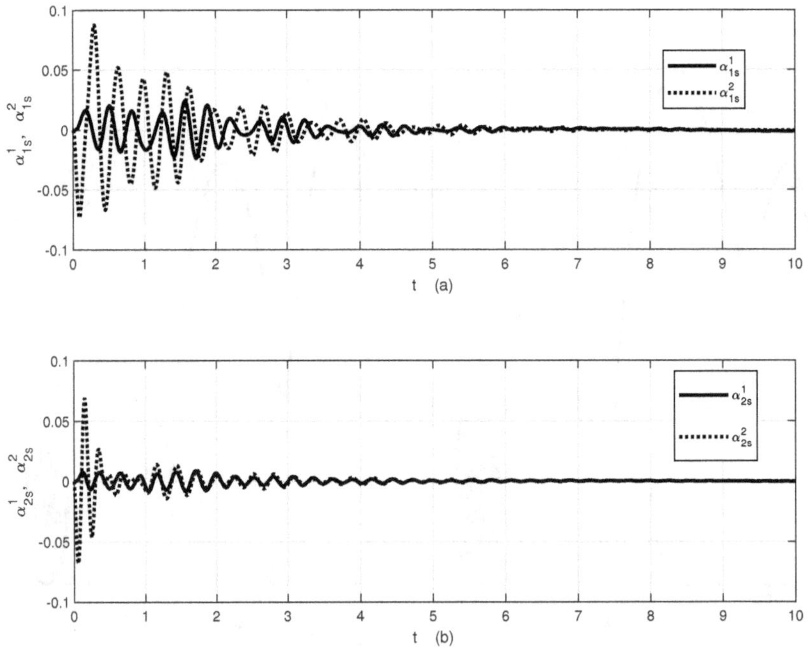

Figure 6.28: Link deflections of the slave manipulators during synchronisation when θ_d is (6.39) with $\beta^1 = 0.5$, $\beta^2 = 0.25$: (a) link-1 and (b) link-2.

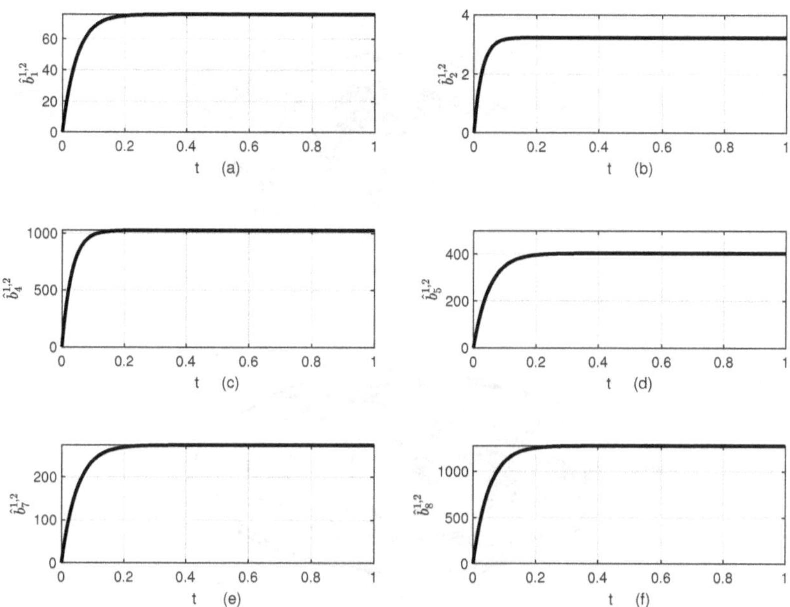

Figure 6.29: Time responses of the estimated parameters of the slave manipulators with $\beta^1 = 0.5, \beta^2 = 0.25$. $\hat{b}_i^{1,2}$ means $\hat{b}_i$ for slave-1 and slave-2.

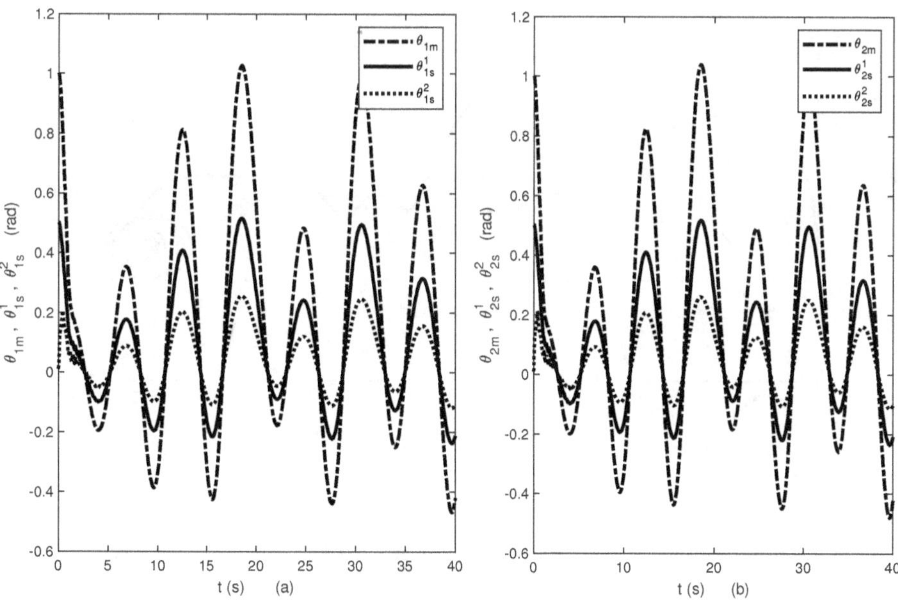

Figure 6.30: Projective synchronisation between the master and slave manipulators in the presence of 0.45 kg payload (with $\beta^1 = 0.5, \beta^2 = 0.25$): (a) link-1 and (b) link-2.

apparent from Fig. 6.31 and Fig. 6.32 that both the slave manipulators synchronised properly with the controlled master manipulator in the presence of the payload. Time responses of the sliding surfaces and control inputs for the slave manipulators during the chaotic trajectory synchronisation in

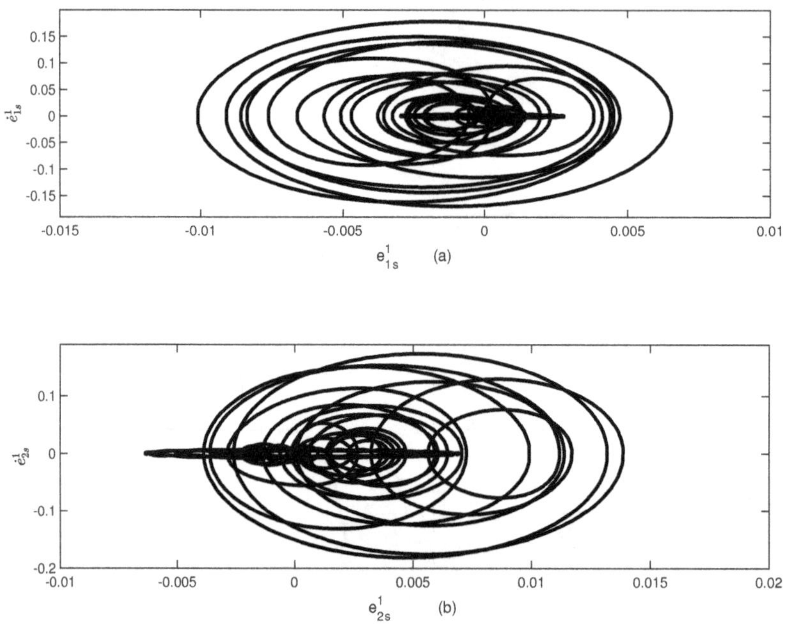

Figure 6.31: Phase plane of the synchronisation errors between the master and slave-1 in the presence of payload: (a) on $e_{1s}^1 - \dot{e}_{1s}^1$ and (b) on $e_{2s}^1 - \dot{e}_{2s}^1$.

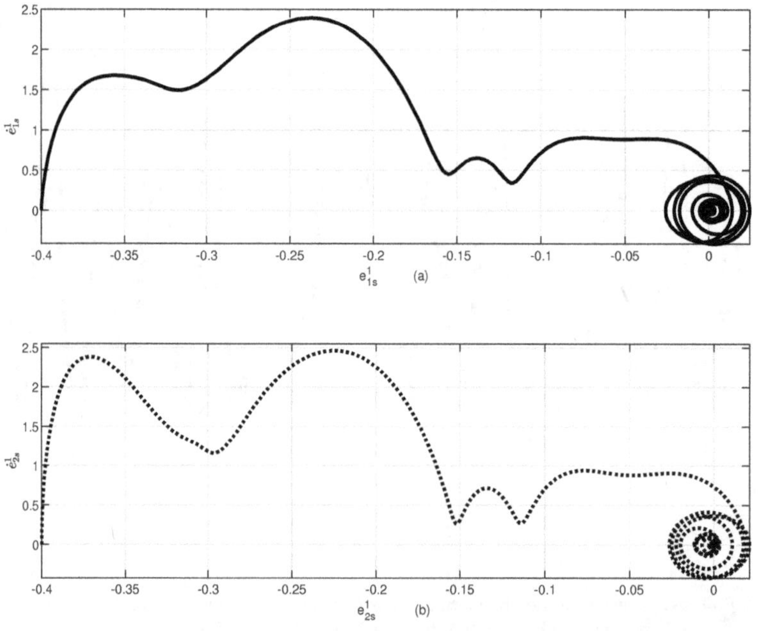

Figure 6.32: Phase plane of the synchronisation errors between the master and slave-2 manipulator in the presence of payload: (a) on $e_{1s}^2 - \dot{e}_{1s}^2$ and (b) on $e_{2s}^2 - \dot{e}_{2s}^2$.

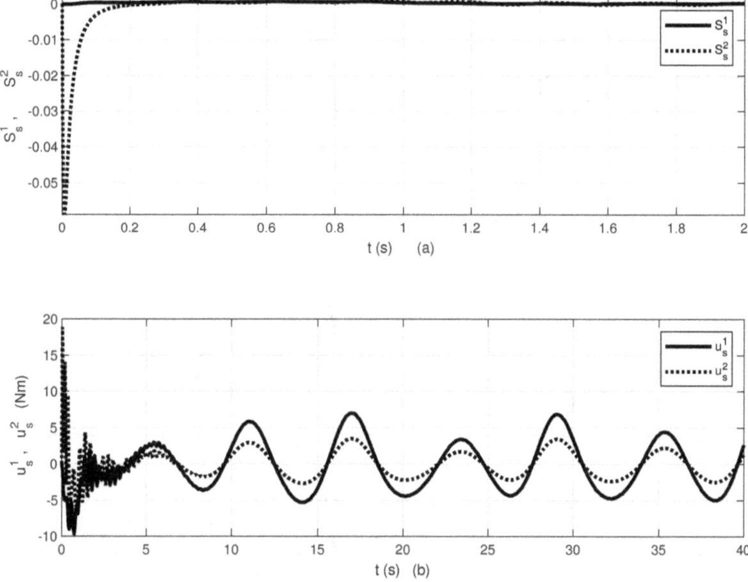

Figure 6.33: (a) Sliding surfaces (S_s^1, S_s^2) and (b) control inputs (u_s^1, u_s^2) for the projective synchronisation in the presence of 0.45 kg payload.

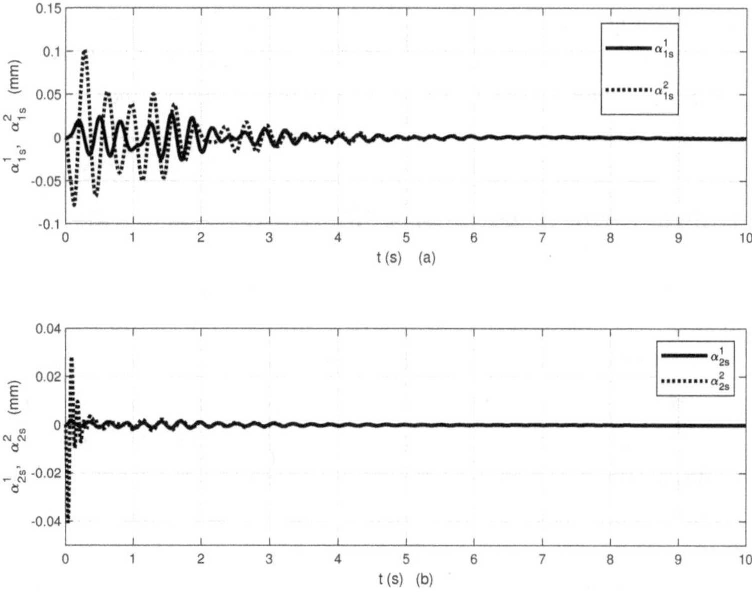

Figure 6.34: Link deflections of the slave manipulators during synchronisation in the presence of payload when θ_d is (6.39) with $\beta^1 = 0.5, \beta^2 = 0.25$: (a) link-1 and (b) link-2.

the presence of payloads are shown in Fig. 6.33. The time responses of the links deflection of both the slave manipulators are shown in Fig. 6.34. It is noted from Fig. 6.34 that the links deflection of both the slave manipulators are suppressed during synchronisation in the presence of payloads, but

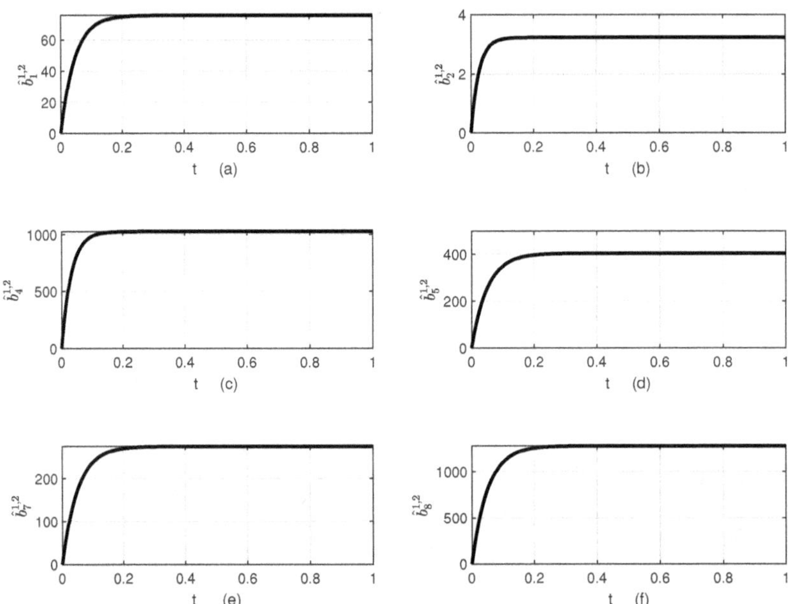

Figure 6.35: Time responses of the estimated parameters of the slave manipulators in the presence of 0.45 kg payload θ_d is (6.39).

deflections are more as compared with or without nominal payloads. Estimation of the parameters of both the slave manipulators are shown in Fig. 6.35. It is observed from Fig. 6.35 that the parameters of the slave manipulators converge to their actual values in the presence of payloads for the chaotic desired trajectory tracking.

6.5.6 Comparison of performances among different variants of SMCs

In this section, the performances of the proposed control technique are compared with the other SMC control techniques. The control techniques are (i) normal SMC with $tanh(s/\epsilon)$ (N-SMC), (ii) adaptive equivalent SMC with $tanh(s/\epsilon)$ (AE-SMC), (iii) modified adaptive equivalent SMC with sigmoid $(s/|s|+\epsilon)$ $(MAE-SMC^1)$ and (iv) modified adaptive equivalent SMC with $tanh(s/\epsilon)$ $(MAE-SMC^2)$. In the case of adaptive SMC, only the parameters of the flexible slave manipulators are considered to be unknown, whereas in case of MAE-SMC, the parameters of the flexible slave manipulators, gain of the sliding surfaces and the of the corrective control law are considered unknowns. The comparisons are done when the master is tracking the desired chaotic signal with a payload of 0.45 kg with the disturbances on the slave manipulators. Figures 6.36 and 6.37 show the responses of the joint angles using the four control techniques for the slave-1 and slave-2 flexible manipulators, respectively. The time responses of the links deflection of the slave-1 and slave-2 manipulators using all the four control technmiques are shown in Figs. 6.38 and 6.39, respectively. Figures 6.40 and 6.41 show the responses of the equivalent sliding surfaces for the slave-1 and slave-2 manipulators, respectively, using the four control techniques. The performance of the controllers for the synchronisation of flexible slave manipulator-1 and slave manipulator-2 are shown in Figs. 6.42 and 6.43, respectively.

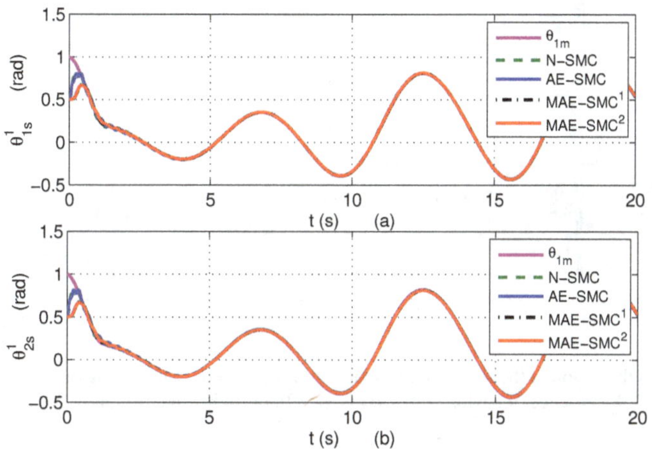

Figure 6.36: Projective synchronisation between the master and slave-1 flexible manipulators using the four controllers in the presence of 0.45 kg payload when θ_d is (6.39).

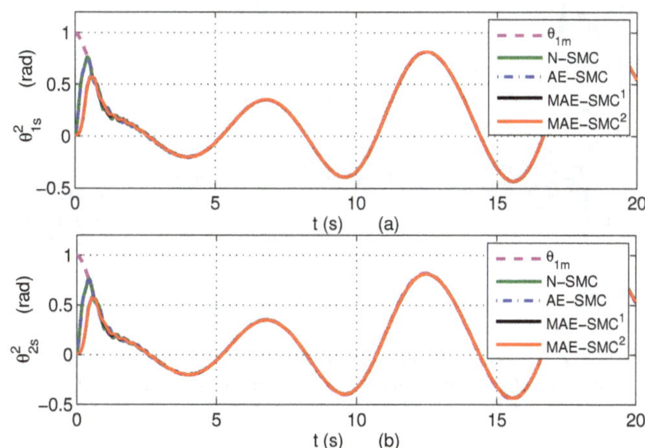

Figure 6.37: Projective synchronisation between the master and slave-2 flexible manipulators using four controllers in the presence of 0.45 kg payload when θ_d is (6.39).

It is seen from Figs. 6.36 and 6.37 that the tracking time for the N-SMC and AE-SMC is less compared with the same of the $(MAE - SMC^1)$ and $(MAE - SMC^2)$ but the steady state error is more in N-SMC and AE-SMC in comparison with $(MAE - SMC^1)$ and $(MAE - SMC^2)$. Figures 6.38 and 6.39 depict that the links deflection for the first link of both the slave manipulators in $(MAE - SMC^2)$ is lower compared with the other three control techniques, whereas, the link deflections for the second link of both the slave manipulators using $(MAE - SMC^1)$ are lower compared with (AE-SMC) and $(MAE - SMC^1)$ but more in comparison with N-SMC.

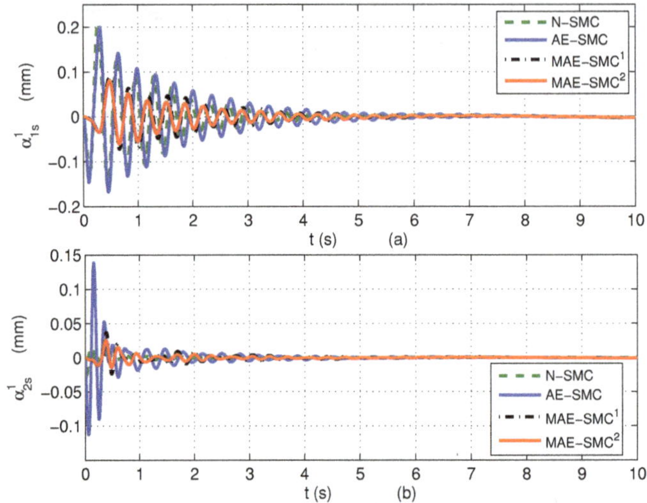

Figure 6.38: Links deflection of the slave-1 manipulator during synchronisation using the four controllers in the presence of payload when θ_d is (6.39): (a) link-1 and (b) link-2.

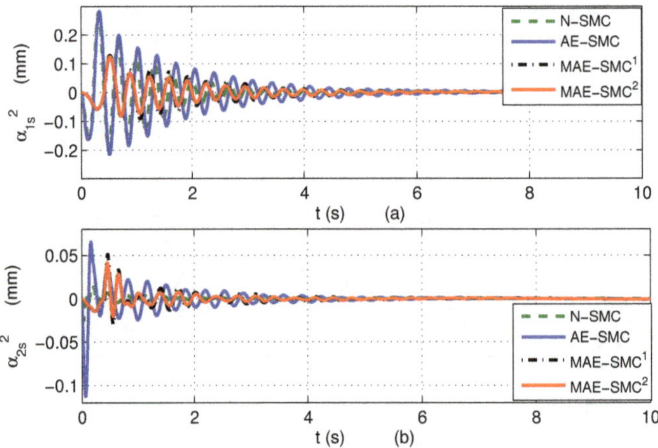

Figure 6.39: Links deflection of the slave-2 manipulator during synchronisation using the four controllers in the presence of payload when θ_d is (6.39): (a) link-1 and (b) link-2.

However, the initial deflection using N-SMC is higher compared with the $(MAE - SMC^2)$. It is apparent from Figs. 6.42 and 6.43 that the required control torque for the synchronisation of both the slave manipulators using $(MAE - SMC^2)$ is very low compared with N-SMC and AE-SMC and slightly lower compared to $(MAE - SMC^1)$. The comparison on the performances of the controllers are given in Table 6.4. The 2-norm and integral square error are the two performance indices considered for comparison. It is seen from the Table 6.4 that the proposed control technique $(MAE - SMC^2)$ requires less control efforts and has less ISE in the case of links deflection for both the slave manipulators.

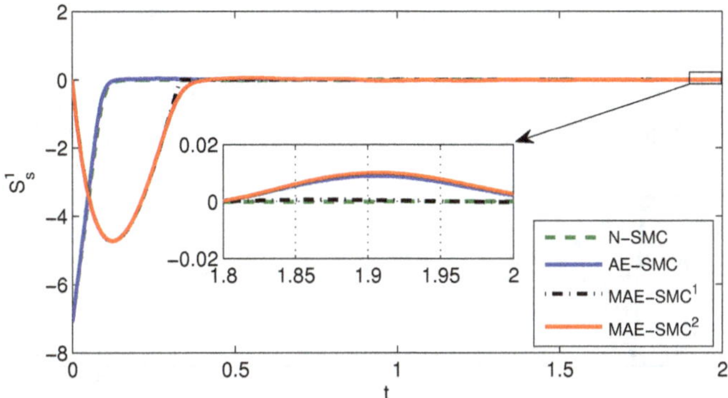

Figure 6.40: (a) Sliding surfaces (S_s^1) for the projective synchronisation of the slave-1 manipulator when θ_d is (6.39) using the four controllers.

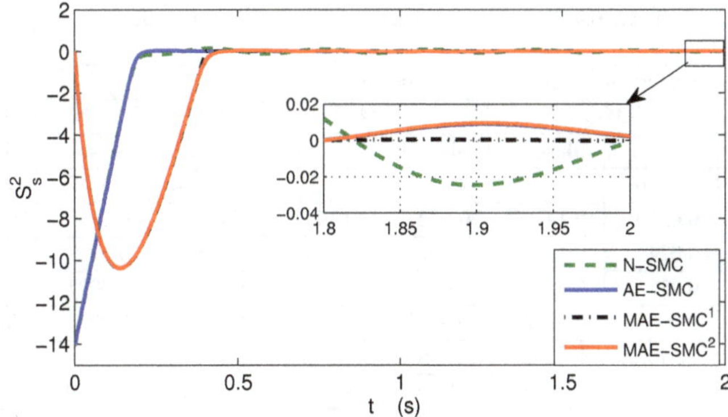

Figure 6.41: (a) Sliding surfaces (S_s^2) for projective synchronisation of the slave-2 manipulator when θ_d is (6.39) using the four controllers.

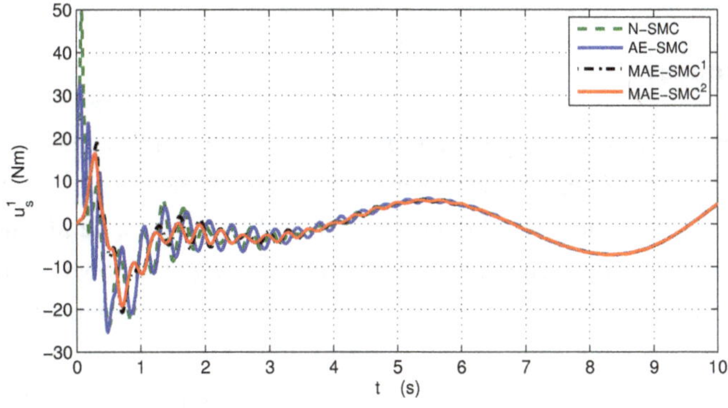

Figure 6.42: Control torque inputs for slave-1 manipulator during synchronisation using the four controllers in the presence of payload when θ_d is (6.39): (a) link-1 and (b) link-2.

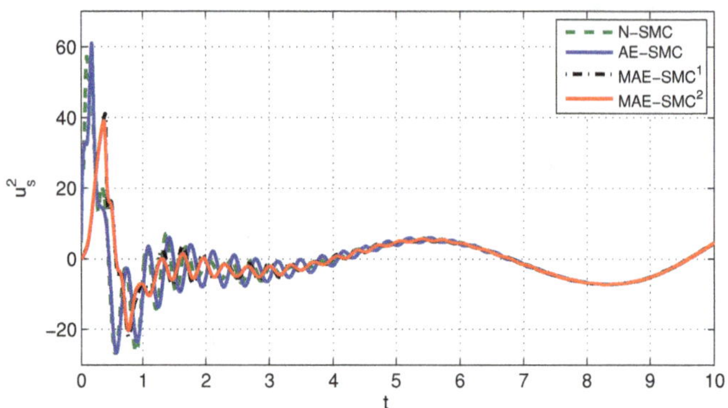

Figure 6.43: Control torque inputs for slave-2 manipulator during synchronisation using the four controllers in the presence of payload when θ_d is (6.39): (a) link-1 and (b) link-2.

Table 6.4: Comparison on performances of the controllers.

	Performances of controllers when θ_d is (6.39)							
Control techniques	$N-SMC$		$AE-SMC$		$MAE-SMC^1$		$MAE-SMC^2$	
Performance Indices	Slave-1	Slave-2	Slave-1	Slave-2	Slave-1	Slave-2	Slave-1	Slave-2
Control energy ($\|\|u\|\|_2$)	1088.6	1241.5	1064.5	1203.2	990.061	1092.2	**981.97**	**1081.7**
ISE of θ_1	**0.0253**	**0.14**	0.026	**0.14**	0.041	0.21	0.041	0.19
ISE of θ_2	0.024	**0.13**	**0.022**	0.121	0.049	0.21	0.040	0.19
ISE of α_1	**0.015**	**0.023**	0.016	0.031	0.0042	0.0071	0.0025	0.0050
ISE of α_2	0.00015	**0.0002**	0.0025	0.0021	0.0003	0.0004	**0.00010**	0.0003
	Performances of controllers when θ_d is (6.38)							
Control energy ($\|\|u\|\|_2$)	570.852	910.047	510.998	842.781	398.313	734.50	**375.885**	**717.999**
ISE of θ_1	0.0025	**0.014**	0.0025	0.015	**0.01**	0.034	0.0058	0.025
ISE of θ_2	0.0025	**0.013**	0.0028	0.014	**0.01**	0.311	0.0058	0.026
ISE of α_1	0.01	0.015	0.016	0.03	0.0040	0.0007	**0.0018**	**0.002**
ISE of α_2	**0.0001**	**0.0001**	0.0004	0.0003	**0.0001**	0.0003	**0.0001**	**0.0001**

6.6 Chapter summary

This chapter attempts the projective synchronisation of multiple identical slave TLFMs and a non-identical controlled master TLFM. This chapter also presents the development, simulation and comparison performances of the four sliding mode control techniques, i.e., (i) normal SMC with $tanh(s/\epsilon)$ (N-SMC), (ii) adaptive equivalent SMC with $tanh(s/\epsilon)$ (AE-SMC), (iii) modified adaptive equivalent SMC with sigmoid $(s/|s|+\epsilon)$ $(MAE-SMC^1)$ and (iv) modified adaptive equivalent SMC with $tanh(s/\epsilon)$ $(MAE-SMC^2)$ for tip trajectory tracking and suppression of tip deflection of TLFMs during projective synchronisation when subjected to variable payloads using desired signals. The performances have been obtained after successfully simulating these control techniques on TLFMs. The simulation results reveal that with changes in the payloads, modified adaptive equivalent SMC with $tanh(s/\epsilon)$ $(MAE-SMC^2)$ exhibits superior performance compared with the other techniques.

References

[1] H. Dou and S. Wang. A boundary control for motion synchronization of a two-manipulator system with a flexible beam. *Automatica*, vol. 50, no. 12, pp. 3088–3099, 2014.

[2] L. D. Khoa, D. Q. Truong, and K. K. Ahn. Synchronization controller for a 3-R planar parallel pneumatic artificial muscle (PAM) robot using modified ANFIS algorithm. *Mechatronics*, vol. 23, no. 4, pp. 462–479, 2013.

[3] S. K. Pradhan and B. Subudhi. Nonlinear Adaptive Model Predictive Controller for a Flexible Manipulator: An Experimental Study. *IEEE Transactions on Control Systems Technology*, vol. 22, no. 5, pp. 1754–1768, 2014.

[4] Z. Mohamed, M. Khairudin, A. R. Husain, and B. Subudhi. Linear Matrix Inequality-based Robust Proportional Derivative Control of a Two-link Flexible Manipulator. *Journal of Vibration and Control*, no. June, pp. 1–12, 2013.

[5] S. K. Pradhan and B. Subudhi. Real-Time Adaptive Control of a Flexible Manipulator Using reinforcement learning. *IEEE Transactions on Automation Science and Engineering*, vol. 9, no. 2, pp. 237–249, 2012.

[6] C. Zhong, Y. Guo, Z. Yu, L. Wang, and Q. Chen. Finite-time Attitude Control for Flexible Spacecraft with Unknown Bounded Disturbance. *Transactions of the Institute of Measurement and Control*, vol. 38, no. 2, pp. 240–249, 2016.

[7] J. T. Agee, Z. Bingul, and S. Kizir. Tip Trajectory Control of a Flexible-link Manipulator using an Intelligent Proportional Integral (iPI) Controller. *Transactions of the Institute of Measurement and Control*, vol. 36, no. 5, pp. 673–682, 2014.

[8] Y. Xiao and K. Y. Zhu. Optimal Synchronization Control of High-Precision Motion Systems. *IEEE Transactions on Industrial Electronics*, vol. 53, no. 4, pp. 1160–1169, 2006.

[9] Y. Su, D. Sun, L. Ren, and J. K. Mills. Integration of saturated PI synchronous control and PD feedback for control of parallel manipulators. *IEEE Transactions on Robotics*, vol. 22, no. 1, pp. 202–207, 2006.

[10] D. Sun. Position Synchronization of Multiple Motion axes with Adaptive Coupling Control. *Automatica*, vol. 39, no. 6, pp. 997–1005, 2003.

[11] Y. C. Liu and N. Chopra. Controlled Synchronization of Heterogeneous Robotic Manipulators in the Task Spaces. *IEEE Transactions on Robotics*, vol. 28, no. 1, pp. 268–275, 2012.

[12] W. Shang, S. Cong, and Y. Ge. Controlled Synchronization of Heterogeneous Robotic Manipulators in the Task Spaces. *IEEE Transactions on Robotics*, vol. 10, no. 3, pp. 1–9, 2010.

[13] D. Zhao, C. Liu, Q. Zhu, and Z. Man. Task Space Synchronized Control for Multiple Robotic Manipulators. In *In: Nanjing: 33rd Chinese Control Conference (CCC), 2014*, pp. 1856 – 1861, 2014.

[14] R. Cui and W. Yan. Mutual synchronization of multiple robot manipulators with unknown dynamics. *Journal of Intelligent and Robotic Systems: Theory and Applications*, vol. 68, no. 2, pp. 105–119, 2012.

[15] P. R. Quyang and V. Pano. Position Domain Synchronization Control of Multi-Degrees of Freedom Robotic Manipulator. *Journal of Dynamic Systems, Measurement, and Control*, vol. 136, no. 2, pp. 021017–021037, 2013.

[16] Z. Li, Y. Xia, and F. Sun. Adaptive Fuzzy Control of Multilateral Cooperative Teleoperation for Multiple Robotic Manipulators under Random Network-induced Delays. *IEEE Transaction On Fuzzy Systems*, vol. 22, no. 2, pp. 437–450, 2014.

[17] L. Chao, Z. Dongya, and X. Xianbo. Force synchronization of multiple robot manipulators: A first study. *Proceedings of the 33rd Chinese Control Conference*, pp. 2212–2217, 2014.

[18] M. S. Soliman and G. Tan. Continuous Query Scheduler Based on Operators Clustering. *Journal of Central South University of Technology*, vol. 15, no. 6, pp. 830–834, 2008.

[19] L. Anic, J. Kasac, and B. Novakovic. Passive sliding-Mode Synchronization of Multi-Robotic Systems with Structural Uncertainties and External Disturbance. In *In: Proceedings of the 22nd International DAAAM Symposium*, pp. 269–270, 2011.

[20] Q. Han. Controlled Synchronization for Master and Slave Manipulators Based on Observed Ender Trajectory. *International Journal of Structural Stability and Dynamics*, vol. 11, no. 6, pp. 1089–1102, 2011.

[21] W. Shang, S. Cong, and S. Jiang. Dynamic Model Based Nonlinear Tracking Control of a Planar Parallel Manipulator. *Nonlinear Dynamics*, vol. 6, no. 4, pp. 597–606, 2011.

[22] H. Dou and S. Wang. Robust adaptive motion/force control for motion synchronization of multiple uncertain two-link manipulators. *Mechanism and Machine Theory*, vol. 67, pp. 77–93, 2013.

[23] Z. Chen, Y. J. Pan, and J. Gu. A Novel Adaptive Robust Control Architecture for Bilateral Teleoperation Systems under Time-Varying Delays. *Int. J. Robust Nonlinear Control*, vol. 25, no. 17, pp. 3349–3366, 2015.

[24] A. Rodriguez-Angeles and H. Nijmeijer. Mutual synchronization of robots via estimated state feedback: A cooperative approach. *IEEE Transactions on Control Systems Technology*, vol. 12, no. 4, pp. 542–554, 2004.

[25] Q. Cao, S. Li, D. Zhao, and Z. Wang. Finite-time motion/force control for motion synchronization of multiple manipulators. *Proceedings of the 33rd Chinese Control Conference, CCC 2014*, pp. 2121–2126, 2014.

[26] M. M. G. Ardakani, J. H. Cho, R. Johansson, and A. Robertsson. Trajectory generation for assembly tasks via bilateral teleoperation. *IFAC Proceedings Volumes*, vol. 47, no. 3, pp. 10230–10235, 2014.

[27] Y. Yang, C. Hua, and X. Guan. Finite Time Control Design for Bilateral Teleoperation System With Position Synchronization Error Constrained. *IEEE Trans. Cybern*, vol. 46, pp. 609–619, 2016.

[28] H. Wang. Passivity based synchronization for networked robotic systems with uncertain kinematics and dynamics. *Automatica*, vol. 49, no. 3, pp. 755–761, 2013.

[29] A. R. Mehrabian and K. Khorasani. Constrained distributed cooperative synchroniszation and reconfigurable control of heterogeneous networked Euler-Lagrange multi-agent systems. *Information Sciences*, vol. 370-371, no. 20, pp. 578–597, 2016.

[30] A. Montano and R. Suarez. Coordination of several robots based on temporal synchronization. *Robotics and Computer-Integrated Manufacturing*, vol. 42, pp. 73–85, number = ,, 2016.

[31] A. Rodriguez-Angeles and H. Nijmeijer. Synchronizing tracking control for flexible joint robots via estimated state feedback. *Journal of Dynamic Systems, Measurement, and Control*, vol. 126, no. 1, pp. 162–172, 2001.

[32] S. Islam, P. X. Liu, and A. E. Saddik. New Stability and Tracking Criteria for A Class of Bilateral Teleoperation Systems. *Inf. Sci*, vol. 278, pp. 868–882, 2014.

[33] S. Islam, P. X. Liu, A. E. Saddik, and Y. B. Yang. Bilateral Control of Teleoperation Systems with Time Delay. *IEEE/ASME Trans. Mechatronics*, vol. 20, pp. 1–12, 2015.

[34] Y. Ye, Y. J. Pan, and T. Hilliard. Bilateral Teleoperation with Time-varying Delay: A Communication Channel Passification Approach. *IEEE/ASME Trans. Mechatronics*, vol. 18, pp. 1431–1434, 2013.

[35] I. Sarras, E. Nuno, and L. Basanez. An Adaptive Controller for Nonlinear Teleoperators with Variable Time-delays. *J. Franklin Inst*, vol. 351, pp. 4817–4837, 2014.

[36] R. Moreau, M. T. Pham, M. Tavakoli, M. Q. Le, and T. Redarce. Sliding-mode Bilateral Teleoperation Control Design for Master-slave Pneumatic Servo Systems. *Control Eng. Pract*, vol. 20, pp. 584–597, 2012.

[37] Y. J. Pan, C. Canudas-de Wit, and O. Sename. A New Predictive Approach for Bilateral Teleoperation with Applications to Drive-by-wire Systems. *IEEE Trans. Robot*, vol. 22, pp. 1146–1162, 2012.

[38] Y. Yang, C. Hua, and X. Guan. Adaptive Fuzzy Finite-time Coordination Control for Networked Nonlinear Bilateral Teleoperation System. *IEEE Trans. Fuzzy Syst*, vol. 22, pp. 631–641, 2014.

[39] M. Ghasemi and S. G. Nersesov. Finite-time Coordination in Multiagent Systems using Sliding Mode Control Approach. *Automatica*, vol. 50, pp. 1209–1216, 2014.

[40] C. Hua, J. Li, Y. Yang, and X. Guan. Extended-state-observer-based finite-time Synchronization Control Design of Teleoperation with Experimental Validation. *Nonlinear Dyn*, vol. 85, pp. 317–331, 2016.

[41] M. Zeinali and L. Notash. Adaptive Sliding Mode Control with Uncertainty Estimator for Robot Manipulators. *Nonlinear Dyn*, vol. 45, pp. 80–90, 2010.

[42] K. H. K, *Nonlinear Systems*. Prentice Hall, 3rd ed. ed., 1970.

[43] G. H. Li. Generalized Projective Synchronization Between Lorenz System and Chen's System. *Chaos, Solitons & Fractals*, vol. 32, no. 4, pp. 1454–1458, 2007.

[44] D. S. Rana. Mathematical Modelling, stability Analysis and control of Flexible Double Link Robotic Manipulator: A Simulation Approach. *IOSR Journal of Engineering*, vol. 3, pp. 29–40, 2013.

[45] G. R and T. R. Harmonic Balance Methods for the Analysis of Chaotic Dynamics in Nonlinear Systems. *Automatica*, vol. 28, pp. 531–548, 1992.

Chapter 7

Synchronisation between Assumed Modes Modelled TLFMs

The control of a TLFM is challenging due to the spatially distributed states which make the system dynamics non-minimum phase and underactuated [1]. The effective operation of serial manipulators require accurate and coordinated control algorithms. Coordination, cooperation and synchronisation are the synonyms used to define the mutual time cooperation between two processes [2]. However, some of the inherent disadvantages [3, 4] of rigid manipulators limit the efficient applications of the serial flexible manipulators. Thus, serial manipulators with coordinated flexible links are more desirable. Many papers are available on the synchronisation of rigid robot manipulators [5], [6], [7–9]. The synchronisation of robot manipulators can be broadly divided into three categories. These are motion synchronisation [10], [2], [9], force synchronisation [7], [9], [11] and task space/trajectory synchronisation [12, 13]. The synchronisation of robot manipulators are categorically explained in [2], [8]. It is to be noted that most of the available synchronisation concepts are focused on the rigid manipulators which are generally not precise, economic and effective in industrial applications. The synchronisation between two-link flexible manipulators (TLFM) is a new approach in industrial applications. The synchronisation between two two-link flexible manipulators is rarely available in the literature.

It is noted from Chapter 4 that the modelling of a flexible manipulator plays an important role while developing the controller for a required operation. It is also notable that the AMM is mostly used and accurate in the modeling method. In this chapter two AMM modelled TLFMs are used for the synchronisation. In the previous chapter (Chapter 6) synchronisation between LPM modeled TLFMs was presented. This chapter exploits the synchronisation between one controlled master and one slave TLFM using AMM modelling. A second order PID terminal SMC (SO-PID-TSMC) control technique is proposed for the synchronisation strategy and for the suppression of tip deflections.

To check the robustness of the synchronisation strategy, external bounded disturbances as well as parametric uncertainties in the range of $\pm 5\%$ to $\pm 30\%$ are added to the master and slave manipulators. An additional payload of 0.155 kg is added to the slave manipulator during synchronisation. The performances of the proposed control technique are compared with those of the normal second order SMC developed in the literature. The chapter is organised as follows. Section 7.2 presents the chapter objectives. Section 7.3 proposes the design of the conventional sliding mode controller for the master manipulator. The design of the proposed second order PID terminal sliding mode controller for synchronisation is presented in Section 7.4. The proposed controllers are applied to the mathematical models of the TLFMs and the simulation results are presented in Section 7.5. The summary of the chapter is presented in Section 7.6.

7.1 Introduction

The important aspects (operation precision) of a synchronisation strategy depend on the type of control algorithms used [14], [15]. Many control techniques are used for the synchronisation of rigid robot manipulators such as adaptive control [8], observer based control, feedback control [2], feedback with PD and saturated PI [16] and computed torque control [17] among others. But, the performance of a control technique deteriorates when uncertainties like modelling errors, backlash, friction and external disturbances [6] are present. The robot manipulator dynamics are generally involved with two types of uncertainties: structured and unstructured uncertainties [18]. The structured uncertainty includes payload variations. Sensor noise, external disturbance, friction and high frequency signals are the unstructured uncertainties [18]. Sliding mode control (SMC) techniques give better results even in the presence of such unstructured uncertainties. Generally, SMC ensures global stability and convergence of system dynamics [19], [20], [21].

Several SMC techniques are used for synchronisation of rigid manipulators like SMC with adaptive control [16], adaptive backstepping SMC [22], low pass filter based SMC [23], integral SMC [6], SMC with fuzzy logic and H_∞. Normally, in a conventional SMC, chattering, high gain and asymptotic stability [19] are considered as disadvantages. Nevertheless, in the first order SMC, chattering can be reduced by using the saturation function in place of the signum function. But, it is seen that the tracking precision and disturbance rejection property get degraded due to the boundedness of the sliding mode variables [19]. An integral SMC is proposed in [6] to overcome the high gain requirement. The terminal sliding surfaces are designed to achieve finite time stability [19], [24], [25]. Terminal sliding mode control (TSMC) techniques are also used for different control problems of a TLFM [26], [27], [28], [29]. However, if the initial states of the system are far away from the equilibrium point, a good convergence performance may not be achieved by a TSMC. Another disadvantage of the first order TSMC is the singularity problem which requires a large control effort [18].

Many methods are available to reduce or to eliminate the chattering such as quasi-SMC, low pass filters, Fuzzy-SMC and higher order SMC (HOSMC) [18]. A HOSMC is considered as a good controller in all respects. In the case of a second order SMC, the actual control is a continuous integration of its derivative owing to the elimination of chattering [19]. Hence, the chattering phenomena are overcome with higher order SMC or other SMC techniques [20].

Motivation:

- It is observed from Table 6.1 that various synchronisations are conducted for the rigid manipulators but none for those of assumed modes method models.
- It is also seen from Table 5.1 (SMC techniques used from TLFMs) that second order sliding mode controller is not designed for the two-link flexible manipulator with parametric uncertainties to check the robustness of a designed controller. The synchronisation of flexible manipulators is expected to be robust with variable payloads and external disturbances. Notably due to the sudden change in payload and external disturbances, uncertainties may occur in the system during synchronisation. Thus, better (SMC) control techniques are required to work well under such circumstances. Thus, the contribution of this chapter lies in developing a new second order PID terminal sliding mode control for a TLFM.

7.2 Chapter objectives

To address the challenges presented in this chapter, the following objectives have been established:

- The first objective is to design a conventional sliding mode controller for the master manipulator, enabling accurate tracking of a desired exponentially varying signal.
- The second objective involves designing a second order sliding mode controller to achieve synchronisation between the controlled master and an identical slave master.
- The third objective is to design the above controller in the presence of payloads and parametric uncertainties between $\pm 5\% to \pm 30\%$.
- The fourth objective is to compare the performances of the designed controllers with the normal second order sliding mode and the proposed controllers.

7.3 Design of a conventional SMC for tracking control of the master manipulator

The dynamics of the master manipulator are as in (3.48) and can be rewritten as

$$B_{0m}(q_m)\ddot{q_m} + h_{0m}(q_m, \dot{q}_m) + K_{0m}q_m + D_{0m}\dot{q}_m = b\tau_m + b\tau_{dm} + Q_m(q, \dot{q}, \ddot{q}) \tag{7.1}$$

The tip position of the master manipulator can be considered as

$$y_{im} = \theta_{im} + \frac{1}{l_{im}} \sum_{j=1}^{2} \phi_{ijm} \delta_{ijm}, \quad i = 1, 2 \tag{7.2}$$

Let y_{di} be the twice differentiable desired tip trajectory for the master manipulator (7.1). The control objective is to track the desired tip trajectory for global stability starting from an arbitrary initial state along with tip deflection suppression.

The proposed synchronisation is achieved in two steps. In the first step, the master manipulator is synchronised with the desired trajectory. In the second step, the slave manipulator is synchronised with the controlled master manipulator. The tracking error for the ith link of the master manipulator is defined as

$$e_{im} = y_{im} - y_{di} \tag{7.3}$$

Now, the objective is to design a suitable sliding mode controller to satisfy the tracking error condition (7.4).

$$\lim_{t \to t_{rm1}, t_{rm2}} \parallel e_{im} \parallel \to 0 \tag{7.4}$$

t_{rm1}, t_{rm2} are the reaching time instants of the sliding surfaces for the master manipulator. The dynamics of the slave manipulator used for the synchronisation is defined in (7.5).

$$B_{0s}(q_s)\ddot{q}_s + h_{0s}(q_s, \dot{q}_s) + K_{0s}q_s + D_s\dot{q}_s = b\tau_s + b\tau_{ds} + Q_s(q, \dot{q}, \ddot{q}) \tag{7.5}$$

The tip position of the i_{th} link of the slave manipulator is described as

$$y_{is} = \theta_{is} + \frac{1}{l_{is}} \sum_{j=1}^{2} \phi_{ijs} \delta_{ijs}, \quad i = 1, 2 \tag{7.6}$$

The synchronisation between the controlled master and slave manipulators depends on the tip trajectory of the master manipulator (7.1). Hence, the tip trajectory synchronisation error for the ith link is defined as

$$e_{is} = y_{is} - y_{im} \tag{7.7}$$

Now, the goal is to design a suitable PID sliding surface type second order terminal SMC so that the resulting synchronisation error satisfies (7.8).

$$\lim_{t \to t_{rsi}} \parallel e_{is} \parallel \to 0 \tag{7.8}$$

t_{rsi}, $i = 1, 2$ is the reaching time of the sliding surface during synchronisation. This is achieved in two steps. Firstly, a linear conventional sliding surface is defined in terms of the synchronisation error. Secondly, a PID type sliding surface is developed for the second order terminal SMC using the linear SMC. The complete synchronisation strategy is shown in Fig. 7.1. The main goal is to

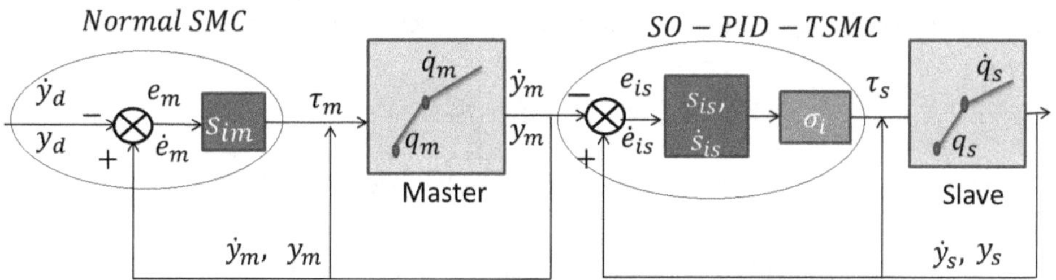

Figure 7.1: The schematic representation of a complete synchronisation strategy between the controlled master and slave manipulators.

synchronise the tip trajectory motion of the slave manipulator with that of the controlled master manipulator.

The inverse dynamic equations of the master manipulator (7.1) can be written as

$$\ddot{q}_m = B_{0m}^{-1}(q_m)[-h_{0m}(q_m,\dot{q}_m) - K_{0m}q_m - D_{0m}\dot{q}_m + b\tau_m + b\tau_{dm} + Q_m(q,\dot{q},\ddot{q})] \tag{7.9}$$

where $q_m = [\theta_{1m}, \theta_{2m}, \delta_{11m}, \delta_{12m}, \delta_{21m}, \delta_{22m}]^T$, and τ_m is the required control input torque for the master manipulator.

The upper boundedness of the system uncertainty in (7.9) is defined as

$$F_m(q,\dot{q},\ddot{q}) = B_{0m}^{-1}(q_m)[b\tau_{dm} + Q_m(q,\dot{q},\ddot{q})] \tag{7.10}$$

which can be considered as

$$\| F_m \| < \gamma_{00} + \gamma_{11} \| q \| + \gamma_{12} \| \dot{q} \|^2 \leq \gamma_3 \tag{7.11}$$

where γ_{00}, γ_{11}, γ_{12} and γ_3 are the positive definite constant matrices. The total displacement for both the links of the master manipulator (7.9) can be written as

$$y_m = \sum_{i=1}^{2} l_{im}\theta_{im} + u_{im} \tag{7.12}$$

where $u_{im} = \sum_{j=1}^{2} \phi_{ijm}\delta_{ijm}$. Thus using (7.12), we can describe the tip position of the master manipulator as in (7.13).

$$y_{im} = \theta_{im} + \frac{1}{l_{im}}(\phi_{i1m}\delta_{i1m} + \phi_{i2m}\delta_{i2m}) \tag{7.13}$$

where i=1, 2. y_{1m} is the tip position of the first link and y_{2m} is the tip position of the second link for the master manipulator (7.1). Let y_{di} be the twice differentiable desired tip trajectories for the links. The expressions of y_{di} are given in the results and discussion section. The tracking errors are defined in (7.14).

$$e_{im} = y_{im} - y_{di} \tag{7.14}$$

The second derivative of (7.14) for both the links can be written as (7.15) and (7.16).

$$\ddot{e}_{1m} = \ddot{q}_m(1) + \frac{1}{l_{1m}}\{\phi_{11m}\ddot{q}_m(3) + \phi_{12m}\ddot{q}_m(4)\} - \ddot{y}_{d1} \tag{7.15}$$

$$\ddot{e}_{2m} = \ddot{q}_m(2) + \frac{1}{l_{2m}}\{\phi_{21m}\ddot{q}_m(5) + \phi_{22m}\ddot{q}_m(6)\} - \ddot{y}_{d2} \tag{7.16}$$

where $\ddot{q}_m(1)$, $\ddot{q}_m(2)$ are the first and second elements of $\ddot{q}_m$, respectively, and so on. The next step is the use of a SMC to design τ_m to stabilise the error dynamics to zero. In the first step of designing the SMC, sliding surfaces are developed which are given in (7.17).

$$s_{im} = \dot{e}_{im} + c_{im}e_{im} \tag{7.17}$$

where $c_{im} > 0$ are user defined constant gains. The necessary and sufficient conditions for the existence of the sliding mode is to satisfy $s_{im} = \dot{s}_{im} = 0$. When the system (7.9) reaches the sliding surface $s_{im} = 0$, the equivalent sliding mode dynamics can be written as:

$$\ddot{e}_{im} = -c_{im}\dot{e}_{im} \tag{7.18}$$

Theorem 7.1. *The sliding mode dynamics of (7.17) are asymptotically stable and converge to the sliding surfaces $s_{im} = \dot{s}_{im} = 0$, i.e., to the equilibrium point $e_{1m} = e_{2m} = 0$.*

Proof. Consider a Lyapunov function candidate.

$$V_{1m}(\dot{e}_{1m}, \dot{e}_{2m}) = \frac{1}{2}(\dot{e}_{1m}^2 + \dot{e}_{2m}^2) \tag{7.19}$$

Taking the time derivative of (7.19) and using (7.18), we get,

$$\dot{V}_{1m}(\dot{e}_{im}) = -(c_{1m}\dot{e}_{1m}^2 + c_{2m}\dot{e}_{2m}^2) \tag{7.20}$$

It is seen that (7.20) is a negative definite function. According to the Lyapunov stability theory, the sliding mode dynamics of (7.17) are asymptotically stable and the trajectories converge to the equilibrium point $e_{1m} = e_{2m} = 0$. □

The next step is to design a control law to ensure that the system trajectories occur on the sliding surfaces. This control law is designed based on the following theorem.

Theorem 7.2. *The tracking errors in (7.14) are asymptotically stable within the reaching time given in (7.24), if the sliding surfaces are chosen as in (7.17) with the control laws as given in (7.21) and (7.22).*

$$\tau_1 = \frac{L_3}{L_1} - \frac{L_2}{L_1}\tau_2 - \rho_{1m} tanh(s_{1m}) \tag{7.21}$$

$$\begin{aligned} \tau_2 = & \frac{L_3 L_4 - L_1 L_6}{L_2 L_4 - L_1 L_5} - \rho_{2m} tanh(s_{2m}) \\ & + \frac{L_1 L_4}{L_1 L_5 - L_2 L_4}\rho_{1m} tanh(s_{1m}) \end{aligned} \tag{7.22}$$

where

$$\begin{cases} R = N_{0m}(h_{0m} + K_{0m}q + D_{0m}\dot{q} \\ L_1 = N_{11} + \phi'_{11m}N_{31} + \phi'_{12m}N_{41} \\ L_2 = N_{12} + \phi'_{11m}N_{32} + \phi'_{12m}N_{42} \\ L_3 = \ddot{y}_{d1} - c_{1m}\dot{e}_{1m} + R(1) + R(3) + R(4) \\ L_4 = N_{21} + \phi'_{21m}N_{51} + \phi'_{22m}N_{61} \\ L_5 = N_{22} + \phi'_{21m}N_{52} + \phi'_{22m}N_{62} \\ L_6 = \ddot{y}_{d2} - c_{2m}\dot{e}_{2m} + R(2) + R(5) + R(6) \end{cases} \tag{7.23}$$

where $N_{0m} = B_{0m}^{-1}$, $R(i)$ are the elements of R, N_{ijm} are the elements of $N_{0m}(i,j)$ and $\phi'_{ijm} = \frac{1}{l_i}(\phi_{ijm})$

$$t_{rm1} = -\frac{|s_{1m}(0)|}{\rho_{1m}}, t_{rm2} = -\frac{|s_{2m}(0)|}{\rho_{2m}} \tag{7.24}$$

where *tanh* is the tangent hyperbolic function, $s_{1m}(0), s_{2m}(0)$ are the initial states of s_{1m} and s_{2m}, respectively. $\rho_{1m}, \rho_{2m} > 0$ are the design parameters. The terms $\ddot{q}^1_m(1), \ddot{q}^1_m(2), \ddot{q}^1_m(3), \ddot{q}^1_m(4), \ddot{q}^1_m(5), \ddot{q}^1_m(6)$ are the elements of the vector $(\ddot{q}^1_m)$ given below

$$\ddot{q}^1_m = B_{0m}^{-1}(q_m)[-h_{0m}(q_m, \dot{q}_m) - K_{0m}q_m - D_{0m}\dot{q}_m + Q_m(q, \dot{q}, \ddot{q})] \tag{7.25}$$

Proof. From (7.9), we can segregate the equation as

$$\ddot{q}_m = N_{0m}\tau_m - N_{0m}(h_{0m} + K_{0m}q_m + D_{0m}\dot{q}_m) \tag{7.26}$$

where $R = N_{0m}(h_{0m} + K_{0m}q_m + D_{0m}\dot{q}_m)$ and R_i, $i = 1, ..., 6$ are the elements of R. (7.26) can be presented as

$$\begin{pmatrix} \ddot{q}_m(1) \\ \ddot{q}_m(2) \\ \ddot{q}_m(3) \\ \ddot{q}_m(4) \\ \ddot{q}_m(5) \\ \ddot{q}_m(6) \end{pmatrix} = \begin{pmatrix} N_{11}\tau_1 + N_{12}\tau_2 - R(1) \\ N_{21}\tau_1 + N_{22}\tau_2 - R(2) \\ N_{31}\tau_1 + N_{32}\tau_2 - R(3) \\ N_{41}\tau_1 + N_{42}\tau_2 - R(4) \\ N_{51}\tau_1 + N_{52}\tau_2 - R(5) \\ N_{61}\tau_1 + N_{62}\tau_2 - R(6) \end{pmatrix} \tag{7.27}$$

where N_{i1}, N_{i2}, $i = 1, ..., 6$ are the elements of N_{0m}.

Taking the time derivative of (7.17) and using the notation given in (7.23) along with (7.15) and (7.16), we can obtain the values of τ_1 and τ_2 as given in (7.21) and (7.22). Then, the derivative of s_{1m} and s_{2m} can be written as

$$\begin{cases} \dot{s}_{1m} = L_1\tau_1 + L_2\tau_2 - L_3 \\ \dot{s}_{2m} = L_3\tau_1 + L_4\tau_2 - L_6 \end{cases} \tag{7.28}$$

Stability of the sliding surfaces (s_{1m}, s_{2m}) are shown using the following Lyapunov function candidate

$$V_a(s_{1m}) = \frac{1}{2}s_{1m}^2 \tag{7.29}$$

$$\dot{V}_a(s_{1m}) = s_{1m}\dot{s}_{1m} \tag{7.30}$$

Taking the time derivative of (7.29) and using that of the first sliding surface as given in 7.28 we 7.31 follows as

$$\dot{V}_a(s_{1m}) = s_{1m}[L_1\tau_1 + L_2\tau_2 - L_3] \tag{7.31}$$

Using the expressions of the torques as given in (7.21), (7.22) along with the notations given in (7.23) and putting in (7.31), we get,

$$\begin{aligned} \dot{V}_a(s_{1m}) &= s_{1m}[L_1\{\frac{L_3}{L_1}\} - \frac{L_2}{L_1}\tau_2 - \rho_{1m}\tanh(s_{1m})\} + L_2\tau_2 - L_3] \\ &= s_{1m}[-L_1\rho_{1m}\tanh(s_{1m})] \\ &= -L_1\rho_{1m}|s_{1m}| \\ &\leq -L_1\rho_{1m}|s_{1m}| \end{aligned} \tag{7.32}$$

Similarly,

$$V_b(s_{2m}) = \frac{1}{2}s_{2m}^2 \tag{7.33}$$

$$\dot{V}_b(s_{2m}) = s_{2m}\dot{s}_{2m} \tag{7.34}$$

Using (7.28), we get,

$$\dot{V}_b(s_{2m}) = s_{2m}[L_4\tau_1 + L_5\tau_2 - L_6] \tag{7.35}$$

Putting (7.21) and (7.22) in (7.35), we get

$$\begin{aligned} \dot{V}_b(s_{2m}) &= s_{2m}[L_4\{\frac{L_3}{L_1}\} - \frac{L_2}{L_1}\tau_2 - \rho_{1m}\tanh(s_{1m})\} + L_5\tau_2 - L_6] \\ &= s_{2m}[(\frac{L_4L_3 - L_6L_1}{L_1}) + (\frac{L_5L_1 - L_4L_3}{L_1}) \\ &- L_4\rho_{1m}\tanh(s_{1m})] \end{aligned} \tag{7.36}$$

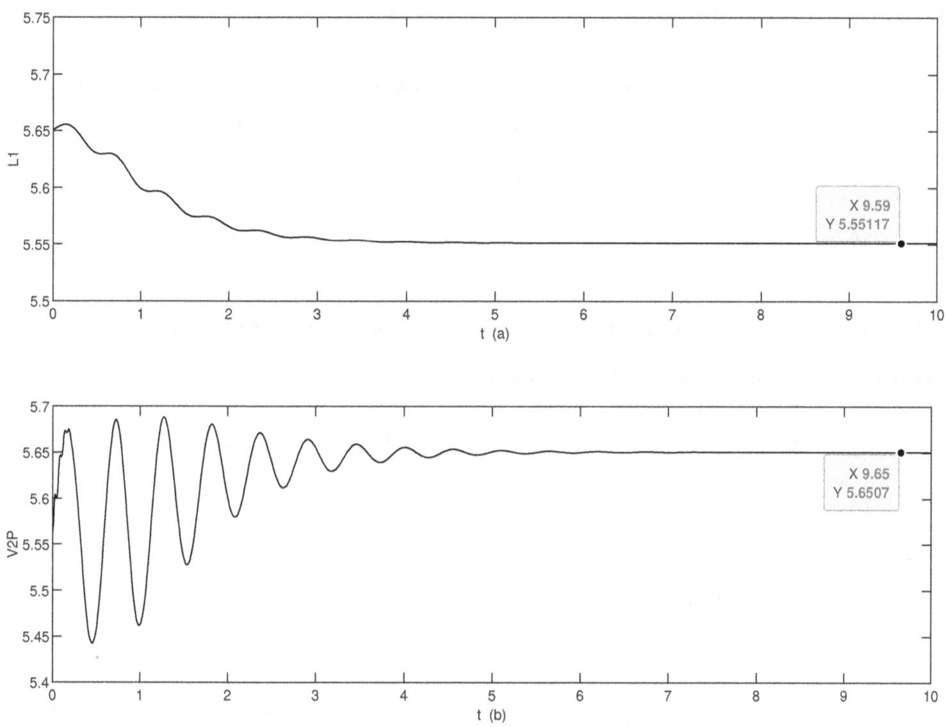

Figure 7.2: The time varying values of L_1 and the term $V2P$.

Putting (7.22) in the (7.36), we get

$$\dot{V}_b(s_{2m}) = s_{2m}[(\frac{L_4L_3 - L_6L_1}{L_1}) + (\frac{L_5L_1 - L_4L_3}{L_1})$$
$$+ (\frac{L_1L_4}{L_1L_5 - L_2L_4})\rho_{1m}\tanh(s_{1m})\} - L_4\rho_{1m}\tanh(s_{1m})] \tag{7.37}$$

$$\dot{V}_b(s_{2m}) = s_{2m}[-(\frac{L_5L_1 - L_2L_4}{L_1})\rho_{2m}\tanh(s_{2m})]$$
$$\dot{V}_b(s_{2m}) \leq -(\frac{L_5L_1 - L_2L_4}{L_1})\rho_{2m}|s_{2m}| \tag{7.38}$$

Here, the notation L_1 and the term $\frac{L_5L_1 - L_2L_4}{L_1} = V2P$ are the time varying positive quantities whose values are depicted in Fig. 7.2. It is seen from Fig. 7.2 that L_1 and the term $\frac{L_5L_1 - L_2L_4}{L_1} = V2P$ are always positive. The positive values of $L1$ and $V2P$ are found to be 5.5 and 5.6, respectively. Hence, the derivative of the Lyapunov function $V_{s_{1m}}$ and $V_{s_{2m}}$ are negative definite. where ρ_{1m}, ρ_{2m}, c_{1m} and c_{2m} are user dependent positive constant gains. The values of ρ_{1m} and ρ_{2m} are considered to be greater than the total boundedness of the elements of the system uncertainty γ_3. Hence, tip trajectories (7.13) of the master manipulator (7.9) follow the desired trajectory within the estimated time given in (7.24). □

Remark 7.3. The parameters ρ_{1m}, ρ_{2m} are crucial for determining the convergence of the sliding surfaces s_{1m}, s_{2m}. The larger of the values of ρ_{1m}, ρ_{2m} will force the system trajectory to converge rapidly on the s_{1m}, s_{2m} (i.e., the tip trajectory of the manipulator dynamics follows the desired trajectory rapidly). However, larger values of ρ_{1m}, ρ_{2m} requires larger control inputs which is not desirable in real practice. Thus, the choices of ρ_{1m}, ρ_{2m} are very crucial and it requires proper adjustment to get a fast response and low control energy consumption. Here, the values of the constant gains ρ_{1m}, ρ_{2m}, c_{1m}, c_{2m} are given in the results and discussion section.

7.4 Design of a second order PID terminal SMC for the synchronisation between the controlled master and slave manipulators

In the previous section, the master manipulator is controlled to follow the desired trajectory. This section discusses the synchronisation of a slave TLFM with the controlled master. The inverse dynamic model of the slave TLFM is given in (7.39).

$$\begin{aligned} \ddot{q}_s =& B_{0s}^{-1}(q_s, \dot{q}_s)[-h_{0s}(q_s, \dot{q}_s) - K_{0s}q_s - D_{0s}\dot{q}_s + b\tau_s \\ &+ b\tau_{ds} + Q_s(q, \dot{q}, \ddot{q})] \end{aligned} \tag{7.39}$$

where $q_s = [\theta_{1s}, \theta_{2s}, \delta_{11s}, \delta_{12s}, \delta_{21s}, \delta_{22s}]^T$ and τ_s is the required control input torque for the slave manipulator. The tip positions of the slave manipulator are,

$$\begin{cases} y_{1s} = \theta_{1s} + \frac{1}{l_{1s}}(\phi_{11s}\delta_{11s} + \phi_{12s}\delta_{12s}) \\ y_{2s} = \theta_{2s} + \frac{1}{l_{2s}}(\phi_{21s}\delta_{21s} + \phi_{22s}\delta_{22s}) \end{cases} \tag{7.40}$$

where $i,\ j = 1,\ 2$, y_{1s} is the tip position for the first link and y_{2s} is the tip position of the second link of the slave manipulator (7.39). The synchronisation error for the master manipulator is the difference between the desired and master trajectories. The difference between the master and slave manipulators is termed as the synchronisation error. Now, the tip trajectory synchronisation errors between the master (7.9) and slave manipulators (7.39) can be written as:

$$e_{is} = y_{is} - y_{im} \tag{7.41}$$

The second derivative of the synchronisation errors in (7.41) can be written as:

$$\begin{aligned} \ddot{e}_{1s} =& -[\ddot{q}_m(1) + \frac{1}{l_{1m}}\{\phi_{11m}\ddot{q}_m(3) + \phi_{12m}\ddot{q}_m(4)\}] \\ &+ [\ddot{q}_s(1) + \frac{1}{l_{1s}}\{\phi_{11s}\ddot{q}_s(3) + \phi_{12s}\ddot{q}_s(4)\}] \end{aligned} \tag{7.42}$$

$$\begin{aligned}\ddot{e}_{2s} &= -[\ddot{q}_m(2) + \frac{1}{l_{2m}}\{\phi_{21m}\ddot{q}_m(5) + \phi_{22m}\ddot{q}_m(6)\}] \\ &+ [\ddot{q}_s(2) + \frac{1}{l_{2s}}\{\phi_{21s}\ddot{q}_s(5) + \phi_{12s}\ddot{q}_s(6)\}]\end{aligned} \tag{7.43}$$

where $\ddot{q}_m(k)$, $\ddot{q}_s(k)$, $(k = 1, ..., 6)$ are the elements of the column vector given in (7.9) and (7.39), respectively. The derivatives of (7.42) and (7.43) are given in (7.44) and (7.45), respectively.

$$\begin{aligned}\frac{d}{dt}(\ddot{e}_{1s}) &= -[\frac{d}{dt}\ddot{q}_m(1) + \frac{1}{l_{1m}}\{\phi_{11m}(\frac{d}{dt}\ddot{q}_m(3)) + \phi_{12m}(\frac{d}{dt}\ddot{q}_m(4))\}] \\ &+ [\frac{d}{dt}\ddot{q}_s(1) + \frac{1}{l_{1s}}\{\phi_{11s}(\frac{d}{dt}\ddot{q}_s(3)) + \phi_{12s}(\frac{d}{dt}\ddot{q}_s(4))\}]\end{aligned} \tag{7.44}$$

$$\begin{aligned}\frac{d}{dt}(\ddot{e}_{2s}) &= -[\frac{d}{dt}\ddot{q}_m(2) + \frac{1}{l_{2m}}\{\phi_{21m}(\frac{d}{dt}\ddot{q}_m(5)) + \phi_{22m}(\frac{d}{dt}\ddot{q}_m(6))\}] \\ &+ [\frac{d}{dt}\ddot{q}_s(2) + \frac{1}{l_{2s}}\{\phi_{21s}(\frac{d}{dt}\ddot{q}_s(5)) + \phi_{22s}(\frac{d}{dt}\ddot{q}_s(6))\}]\end{aligned} \tag{7.45}$$

where $\frac{d}{dt}\ddot{q}_m(k)$, $(k = 1, ...6)$ are the elements of the column vector in (7.46).

$$\begin{aligned}\frac{d}{dt}(\ddot{q}_m) &= \{\dot{B}_{0m}^{-1}(q_m)[-h_{0m}(q_m, \dot{q}_m) - K_{0m}q_m - D_{0m}\dot{q}_m + b\tau_m + b\tau_{dm} \\ &+ Q_m(q, \dot{q}, \ddot{q})]\} + \{B_{0m}^{-1}(q_m)[-\frac{d}{dt}h_{0m}(q_m, \dot{q}_m) - \frac{d}{dt}K_{0m}q_m \\ &- \frac{d}{dt}(D_{0m})\dot{q}_m + b\dot{\tau}_m + b\dot{\tau}_{dm} + \dot{Q}_m(q, \dot{q}, \ddot{q})]\}\end{aligned} \tag{7.46}$$

$$\begin{aligned}&= \{\dot{B}_{0m}^{-1}(q_m)[-h_{0m}(q_m, \dot{q}_m) - K_{0m}q_m - D_{0m}\dot{q}_m + b\tau_m]\} \\ &+ \{B_{0m}^{-1}(q_m)[-\frac{d}{dt}h_{0m}(q_m, \dot{q}_m) - \frac{d}{dt}K_{0m}q_m - \frac{d}{dt}D_{0m}\dot{q}_m + b\dot{\tau}_m]\} \\ &+ \{\dot{B}_{0m}^{-1}(q_m)[b\tau_{dm} + Q_m(q, \dot{q}, \ddot{q})]\} + \{B_{0m}^{-1}(q_m)[b\dot{\tau}_{dm} + \dot{Q}_m(q, \dot{q}, \ddot{q})]\}\end{aligned} \tag{7.47}$$

The $\frac{d}{dt}\ddot{q}_s(k)$, $(k = 1, ...6)$ are the elements of the column vector given in (7.48).

$$\begin{aligned}\frac{d}{dt}(\ddot{q}_s) &= \{\dot{B}_{0s}^{-1}(q_s)[-h_{0s}(q_s, \dot{q}_s) - K_{0s}q_s - D_{0s}\dot{q}_s + b\tau_s \\ &+ b\tau_{ds} + Q_s(q, \dot{q}, \ddot{q})]\} + \{B_{0s}^{-1}(q_s)[-\frac{d}{dt}h_{0s}(q_s, \dot{q}_s) \\ &- \frac{d}{dt}K_{0s}q_s - \frac{d}{dt}(D_{0s})\dot{q}_s + b\dot{\tau}_s + b\dot{\tau}_{ds} + \dot{Q}_s(q, \dot{q}, \ddot{q})]\}\end{aligned} \tag{7.48}$$

$$\begin{aligned}&= \{\dot{B}_{0s}^{-1}(q_s)[-h_{0s}(q_s, \dot{q}_s) - K_{0s}q_s - D_{0s}\dot{q}_s + b\tau_s]\} \\ &+ \{B_{0s}^{-1}(q_s)[-\frac{d}{dt}h_{0s}(q_s, \dot{q}_s) - \frac{d}{dt}K_{0s}q_s - \frac{d}{dt}D_{0s}\dot{q}_s + b\dot{\tau}_s + b\dot{\tau}_{ds}]\} \\ &+ \{\dot{B}_{0s}^{-1}(q_s)[b\tau_{ds} + Q_s(q, \dot{q}, \ddot{q})]\} + \{B_{0s}^{-1}(q_s)[b\dot{\tau}_{ds} + \dot{Q}_s(q, \dot{q}, \ddot{q})]\}\end{aligned} \tag{7.49}$$

The system uncertainties in (7.48) and (7.49) are,

$$\begin{aligned} P_m &= \{\dot{B}_{0m}^{-1}(q_m)[b\tau_{dm} + Q_m(q,\dot{q},\ddot{q})]\} + \{B_{0m}^{-1}(q_m)[b\dot{\tau}_{dm} + \dot{Q}_m(q,\dot{q},\ddot{q})]\} \\ P_s &= \{\dot{B}_{0s}^{-1}(q_s)[b\tau_{ds} + Q_s(q,\dot{q},\ddot{q})]\} + \{B_{0s}^{-1}(q_s)[b\dot{\tau}_{ds} + \dot{Q}_s(q,\dot{q},\ddot{q})]\} \end{aligned} \tag{7.50}$$

The combination of the boundedness of the above uncertainties in the controlled and slave manipulators are,

$$\begin{cases} \| W_m \| = \gamma_{0m} + \gamma_{1m} \| q_m \| + \gamma_{2m} \| \dot{q}_m \| < \gamma_{3m} \\ \| W_s \| = \gamma_{0s} + \gamma_{1s} \| q_s \| + \gamma_{2s} \| \dot{q}_s \| < \gamma_{3s} \end{cases} \tag{7.51}$$

where γ_{0m}, γ_{1m}, γ_{2m}, γ_{3m}, γ_{0s}, γ_{1s}, γ_{2s}, γ_{3s} are the positive constants. *Remark* 7.2. In the case of higher order sliding mode control, i.e., in n^{th} order sliding mode control the condition $s(t) = \dot{s}(t) = \dot{s}^{(n-1)}(t) = 0$ has to be satisfied. Thus (in higher order sliding mode control), the error dynamics are required to force movement to $s(t) = \dot{s}(t) = 0$ while keeping the value of the $(n-1)^{th}$ derivative zero.

Here, the task of the SMC technique is to operate on the slave manipulator (7.39) such that the slave manipulator follows the controlled master manipulator (7.9). The synchronisation of the controlled master and slave manipulators is achieved using the second order proportional integral derivative terminal SMC (SO-PID-TSMC).

The conventional sliding surfaces in terms of the synchronisation errors can be described as:

$$s_{is} = \dot{e}_{is} + c_{is}e_{is} \tag{7.52}$$

where $c_{is} > 0$ are the user dependent constant gains.

The first and second order time derivatives of (7.52) can be written as:

$$\begin{cases} \dot{s}_{is} = \ddot{e}_{is} + c_{is}\dot{e}_{is} \\ \ddot{s}_{is} = \frac{d}{dt}\ddot{e}_{is} + c_{is}\ddot{e}_{is} \end{cases} \tag{7.53}$$

The PID terminal sliding surfaces are designed in terms of the above conventional sliding surfaces defined in (7.54) and (7.55):

$$\sigma_1 = k_{D1}\dot{s}_{1s} + k_{P1}s_{1s} + k_{I1}\int_0^t (tanh(s_{1s})|s_{1s}|^{\alpha_1})d\tau \tag{7.54}$$

$$\sigma_2 = k_{D2}\dot{s}_{2s} + k_{P2}s_{2s} + k_{I2}\int_0^t (tanh(s_{2s})|s_{2s}|^{\alpha_2})d\tau \tag{7.55}$$

where k_{P1}, k_{D1}, k_{I1}, k_{P2}, k_{D2}, k_{I2} are positive gains representing the user dependent design parameters. These gain parameters provide flexibility for the desirable construction of the sliding surfaces. The values of these gains are given in the results and discussion section. Here, the structures of the second order sliding surfaces are in the proportional, integral and derivative (PID) form. The terminal word is referred because the integral part of the sliding surfaces consists of fractional power α which is between 0 to 1. *Remark* 7.3. Here, in PID sliding surfaces ((7.54), (7.55)), it is assumed that the system trajectories are initially in the region $s_i > 0$, i.e., away from the sliding surface. It is also assumed that the control input is not sufficient to drive the error towards the sliding surfaces (since sliding surfaces are a function of errors). Thus, it may cause system trajectories to keep moving away from the sliding surfaces. But the integral action will increase the control input and will be sufficient after some time to drive the trajectories and maintaining them on $s_i = \dot{s}_i = 0$. Once the trajectories approach $\sigma_i = 0$, the control action decreases because s is reducing. The non-singular terms in the σ_i assure finite time convergence and complete error dynamics stability. The conventional sliding surface s_i combines with the proposed SO-PID terminal sliding surface. As σ_i reaches the origin within a finite time boundary, both $s_i, \dot{s}_i$ are bound to converge to the origin.

The necessary and sufficient condition for the existence of sliding mode ((7.54), (7.55)) is $\sigma_i = \dot{\sigma}_i = 0$. Thus, the time derivatives of (7.54) and (7.55) can be written as,

$$\dot{\sigma}_1 = k_{D1}\ddot{s}_{1s} + k_{P1}\dot{s}_{1s} + k_{I1}\{tanh(s_{1s})|s_{1s}|^{\alpha_1}\} \tag{7.56}$$

$$\dot{\sigma}_2 = k_{D2}\ddot{s}_{2s} + k_{P2}\dot{s}_{2s} + k_{I2}\{tanh(s_{2s})|s_{2s}|^{\alpha_2}\} \tag{7.57}$$

Therefore, the equivalent dynamics of the proposed SO-PID-TSMC are,

$$\ddot{s}_{1s} = \frac{1}{k_{D1}}\{-k_{P1}\dot{s}_{1s} - k_{I1}(tanh(s_{1s})|s_{1s}|^{\alpha_1})\} \tag{7.58}$$

$$\ddot{s}_{2s} = \frac{1}{k_{D2}}\{-k_{P2}\dot{s}_{2s} - k_{I2}(tanh(s_{2s})|s_{2s}|^{\alpha_2})\} \tag{7.59}$$

Remark 7.4. The equivalent sliding mode dynamics in (7.58), (7.59) are finite time stable and the trajectories converge to the sliding surfaces $s_{1s} = s_{2s} = 0$ within finite time, using Lemma 1 and 2 given in [30, 31].

Proof. Let, $\dot{s}_{1s} = s_2^+$ and $\dot{s}_{2s} = s_3^+$, then, (7.58) and (7.59) are,

$$\dot{s}_2^+ = -\frac{1}{k_{D1}}\{k_{P1}s_2^+ + k_{I1}(tanh(s_{1s})|s_{1s}|^{\alpha_1})\}$$

$$\dot{s}_3^+ = -\frac{1}{k_{D2}}\{k_{P2}s_3^+ + k_{I2}(tanh(s_{2s})|s_{2s}|^{\alpha_2})\}$$

Consider the following Lyapunov function in the form,

$$V_1(s_{1s}, s_2^+) = \frac{1}{2}(s_{1s})^2 + \frac{1}{2}(s_2^+)^2 \tag{7.60}$$

$$V_2(s_{2s}, s_3^+) = \frac{1}{2}(s_{2s})^2 + \frac{1}{2}(s_3^+)^2 \tag{7.61}$$

Taking the time derivatives of (7.60), (7.61) and using the above expressions, we get,

$$\dot{V}_1(s_{1s}, s_2^+) = s_{1s}s_2^+ - s_2^+[\frac{1}{k_{D1}}\{k_{P1}s_2^+ + k_{I1}(tanh(s_{1s})|s_{1s}|^{\alpha_1})\}] \tag{7.62}$$

$$\dot{V}_2(s_{2s}, s_3^+) = s_{2s}s_3^+ - s_3^+[\frac{1}{k_{D2}}\{k_{P2}s_3^+ + k_{I2}(tanh(s_{2s})|s_{2s}|^{\alpha_2})\}] \tag{7.63}$$

As $tanh(s_{is}) = \frac{s_{is}}{|s_{is}|}$, $(i = 1, 2)$, (7.62)-(7.63) can be simplified to,

$$\dot{V}_1(s_{1s}, s_2^+) = s_{1s}s_2^+ - \frac{k_{P1}}{k_{D1}}(s_2^+)^2 - \frac{k_{I1}}{k_{D1}}s_2^+ s_{1s}|s_{1s}|^{\alpha_1 - 1} \tag{7.64}$$

$$\dot{V}_2(s_{2s}, s_3^+) = s_{2s}s_3^+ - \frac{k_{P2}}{k_{D2}}(s_3^+)^2 - \frac{k_{I2}}{k_{D2}}s_3^+ s_{2s}|s_{2s}|^{\alpha_2 - 1} \tag{7.65}$$

Considering Lemma 1 and Lemma 2 of [32] and assuming $\frac{k_{I1}}{k_{D1}} > 1$, $\frac{k_{I2}}{k_{D2}} > 1$, we can say $\dot{V}_1(s_{1s}, s_2^+)$ and $\dot{V}_2(s_{2s}, s_3^+)$ are negative definite since $0 < \alpha_i < 1$. Hence, the designed SO-PID-TSMC dynamics converge to the sliding surface $s_{1s} = s_{2s} = 0$ within the finite time t_{s1}, t_{s2}. □

Once the trajectories reach the sliding surfaces, they need to be maintained on it using the appropriate control inputs. The structure of the control laws for synchronisation are obtained using Theorem 7.3.

Theorem 7.3. *The master (7.9) and slave (7.39) manipulator synchronise, i.e., the synchronisation errors (7.41) converge to zero within finite time if the second order PID terminal sliding surfaces are chosen as (7.54), (7.55) and the control laws are as follows:*

$$\dot{\tau}_{1s} = \frac{F_9}{F_7} - \frac{F_8}{F_7}\dot{\tau}_{2s} - \rho_{11s}\sigma_1 - \rho_{12s}tanh(\sigma_1) \tag{7.66}$$

$$\begin{aligned}\dot{\tau}_{2s} = & \frac{F_9Z_3 - F_7Z_6}{F_8Z_3 - F_7Z_4} - \rho_{21s}\sigma_2 - \rho_{22s}tanh(\sigma_2) \\ & + (\frac{F_7Z_3}{Z_4F_7 - Z_3F_8})(\rho_{11s}\sigma_1 + \rho_{12s}tanh(\sigma_1))\end{aligned} \tag{7.67}$$

where,

$$\begin{cases} F_1 = K_{I1} tanh(s_{1s}) \mid s_{1s} \mid^{\alpha_1} \\ F_2 = \dot{N}_{0s}(\tau_s + \tau_{ds} - h_{0s}(q_s, \dot{q}_s) - K_{0s}q_s - D_{0s}\dot{q}_s) \\ F_3 = \frac{d}{dt}h_{0s}(q_s, \dot{q}_s) - \frac{d}{dt}K_{0s}q_s - \frac{d}{dt}D_{0s}\dot{q}_s \\ F_4 = F_2 - N_s F_3 \\ F_5 = \frac{d}{dt}\ddot{q}_m(1) + \frac{\phi_{11m}}{l_{1m}}\frac{d}{dt}\ddot{q}_m(3) + \frac{\phi_{12m}}{l_{1m}}\frac{d}{dt}\ddot{q}_m(4) - c_{1s}\ddot{e}_{1s} \\ F_6 = -F_5 + F_4(1) + \phi'_{11s}F_4(3)\phi'_{12s}F_4(4) \\ F_7 = N_{11s} + \phi'_{11s}N_{31s} + \phi'_{12s}N_{41s} \\ F_8 = N_{12s} + \phi'_{11s}N_{32s} + \phi'_{12s}N_{42s} \\ F_9 = \frac{-F_1}{K_{D1}} - F_6 \end{cases} \tag{7.68}$$

Also,

$$\begin{cases} Z_1 = K_{P2}\dot{s}_{2s} + K_{I2} tanh(s_{2s}) \mid s_{2s} \mid^{\alpha_2} \\ Z_2 = \frac{d}{dt}\ddot{q}_m(2) + \phi'_{21m}\frac{d}{dt}\ddot{q}_m(5) + \phi'_{22m}\frac{d}{dt}\ddot{q}_m(6) - c_{2s}\ddot{e}_{2s} \\ Z_3 = N_{21s} + \phi'_{21s}N_{51s} + \phi'_{22s}N_{61s} \\ Z_4 = N_{22s} + \phi'_{21s}N_{52s} + \phi'_{22s}N_{62s} \\ Z_5 = F_4(2) + \phi'_{21s}F_4(5) + \phi'_{21s}F_4(6) - Z_2 \\ Z_6 = -F_5 + F_4(1) + \phi'_{11s}F_4(3)\phi'_{12s}F_4(4) \end{cases} \tag{7.69}$$

$\rho_{11s}, \rho_{12s}, \rho_{22s}, \rho_{22s}$ are positive constant design parameters. $N_{0s} = B_{0s}^{-1}$, N_{ijs} are the elements of $N_{0s}(i, j)$ and $\phi'_{ijs} = \frac{1}{l_i}(\phi_{ijs})$, $\ddot{q}_s(k)$, $(k = 1, ..., 6)$ are the elements of (7.9). The actual control inputs can be obtained by the integration of $\dot{\tau}_{1s}, \dot{\tau}_{2s}$ given in (7.66), (7.67).

Proof. The Lyapunov function candidate is taken as

$$V_{1s}(\sigma_1) = \frac{1}{2}\sigma_1^2 \tag{7.70}$$

$$\dot{V}_{1s}(\sigma_1) = \sigma_1\dot{\sigma}_1 \tag{7.71}$$

Taking the time derivatives of $\dot{s}_{1s}$ and $\dot{s}_{1s}$, we get,

$$\begin{cases} \ddot{s}_{1s} = F_7\dot{\tau}_{s1} + F_8\dot{\tau}_{s2} - F_9 \\ \ddot{s}_{2s} = Z_3\dot{\tau}_{s1} + Z_4\dot{\tau}_{s2} - Z_6 \end{cases} \tag{7.72}$$

Putting these results in the time derivative of (7.54) and using the notation given in (7.68), (7.69), we get,

$$\dot{V}_{1s}(\sigma_1) = \sigma_1[F_7\dot{\tau}_{1s} + F_8\dot{\tau}_{2s} - F_9] \tag{7.73}$$

Putting (7.66) in (7.73), we get

$$\begin{aligned}\dot{V}_{1s}(\sigma_1) =&\sigma_1[F_7\{(\frac{F_9}{F_7}) - (\frac{F_8}{F_7})\dot{\tau}_2 - \rho_{11s}\sigma_1 - \rho_{12s}tanh(\sigma_1)\}\\ &+ F_8\dot{\tau}_{2s} - F_9]\end{aligned} \tag{7.74}$$

$$\begin{aligned}\dot{V}_{1s}(\sigma_1) &= \sigma_1[-F_7(\rho_{11s}\sigma_1 - \rho_{12s}\tanh(\sigma_1))]\\ \dot{V}_{1s}(\sigma_1) &\leq -F_7(\rho_{11}\sigma_1^2 + \rho_{12}|\sigma_1|)\end{aligned} \tag{7.75}$$

Similarly, for the other Lyapunov function candidate

$$V_{2s}(\sigma_2) = \frac{1}{2}\sigma_2^2 \tag{7.76}$$

$$\dot{V}_{2s}(\sigma_2) = \sigma_2\dot{\sigma}_2 \tag{7.77}$$

Taking time derivative of (7.55) and putting the results in (7.77) we get,

$$\dot{V}_{2s}(\sigma_2) = \sigma_2[Z_3\dot{\tau}_{1s} + Z_4\dot{\tau}_{2s} - Z_6] \tag{7.78}$$

Using (7.66) and (7.67), we get

$$\begin{aligned}\dot{V}_{2s}(\sigma_2) &= \sigma_2[Z_3\{(\frac{F_9}{F_7}) - (\frac{F_8}{F_7}) - \rho_{11s}\sigma_1 - \rho_{12s}\tanh(\sigma_1)\}\\ &+ Z_4\dot{\tau}_{2s} - Z_6]\\ &= \sigma_2[(\frac{Z_3F_9 - Z_6F_7}{F_7}) + (\frac{Z_4F_7 - Z_3F_8}{F_7}\dot{\tau}_{2s})\\ &- Z_3(\rho_{11s}\sigma_1 + \rho_{12s}\tanh(\sigma_1)]\end{aligned} \tag{7.79}$$

Putting (7.67) in (7.79), we get,

$$\begin{aligned}\dot{V}_2(\sigma_2) = \sigma_2[&(\frac{Z_3F_9 - Z_6F_7}{F_7}) + (\frac{Z_4F_7 - Z_3F_8}{F_7})\\ &\{(\frac{F_9Z_3 - F_7Z_6}{F_8Z_3 - F_7Z_4}) + (\frac{F_7Z_3}{Z_4F_7 - Z_3F_8})\rho_{11s}\sigma_1\\ &+ (\frac{F_7Z_3}{Z_4F_7 - Z_3F_8})\rho_{12s}\tanh(\sigma_1) - \rho_{21s}\sigma_2 - \rho_{22s}\tanh_{\sigma_2}\}\\ &- Z_3\rho_{11s}\sigma_1 - Z_3\rho_{12s}\tanh(\sigma_1)]\end{aligned} \tag{7.80}$$

$$\begin{aligned}\dot{V}_2(\sigma_2) &= \sigma_2[-(\frac{Z_4F_7 - Z_3F_8}{F_7})(\rho_{21s}\sigma_2 + \rho_{22s}\tanh(\sigma_2))]\\ &\leq -(\frac{Z_4F_7 - Z_3F_8}{F_7})(\rho_{21s}\sigma_2^2 + \rho_{22s}|\sigma_2|)\end{aligned} \tag{7.81}$$

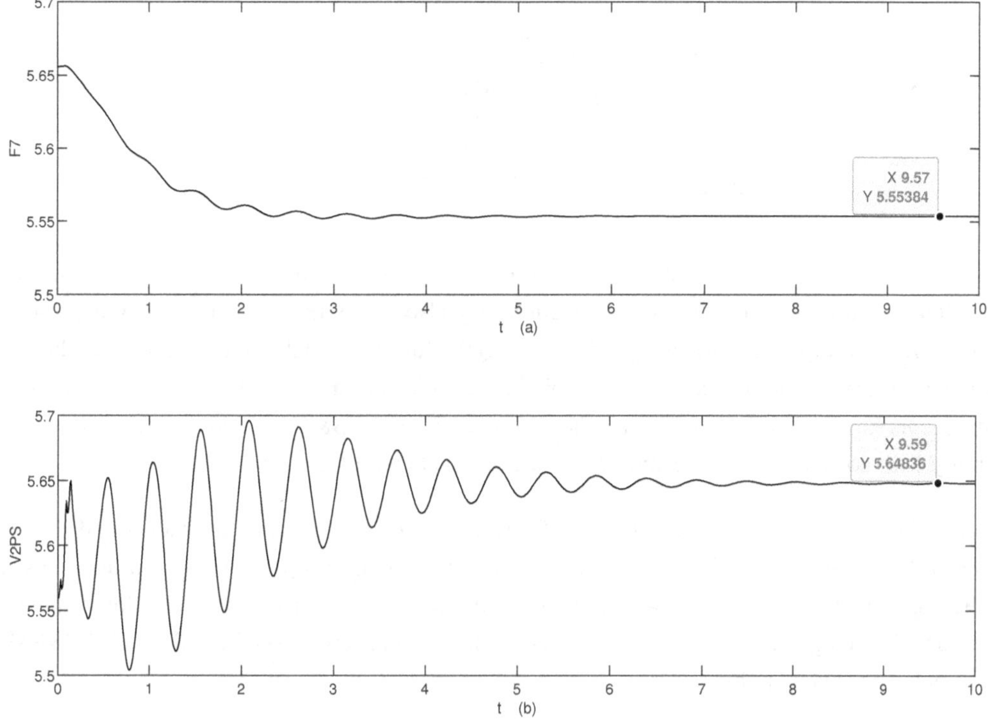

Figure 7.3: The time varying values of F_7 and the term $V2PS$.

Here, the terms F_7 and $\frac{Z_4F_7-Z_3F_8}{F_7} = V2PS$ are the time varying positive quantities whose values are shown in Fig. 7.3. It is apparent from Fig. 7.3 that the positive values of F_7 and the term $\frac{Z_4F_7-Z_3F_8}{F_7} = V2PS$ are found to be 5.5 and 5.6, respectively. Hence, the derivatives of the Lyapunov functions $V_{1s}(\sigma_1)$ and $V_{2s}(\sigma_2)$ are negative definite. Also, ρ_{11s}, ρ_{12s}, ρ_{21s} and ρ_{22s} are the positive gains constants. Here, we consider the values of these positive gains constant to be greater than the total upper boundedness of the system elements uncertainties (γ_{3m}, γ_{3s}). Thus, according to the Lyapunov stability theory, (7.81) and (7.75) are negative definite functions where finite time stability is guaranteed. Hence, the dynamics of the slave manipulator (7.39) follow those of the controlled master (7.9). The tip trajectory synchronisation between the controlled master and the slave manipulators are achieved within finite time, t_{rsi}. □

Suppose, t_{rsi} is the time taken by the sliding surface σ_i to move from $\sigma_i \neq 0$ to $\sigma_i = 0$. The control laws (7.66) and (7.67) help to maintain σ_i at the origin. Thus, the sliding surface s_{is} converges to zero within finite time. The finite time t_{rsi} is determined using (7.82).

$$t_{rsi} \leq \frac{|\sigma_i(0)|}{\rho_{ijs}} \tag{7.82}$$

for $i, j = 1, 2$, where $\sigma_i(0)$ is the initial state of the sliding surface $\sigma_i(t)$.

Table 7.1: Initial conditions used for different equations.

Equation No.	Initial Conditions
(7.1)	$q_m = (0.1, 0.1, 0, 10^{-4}, 10^{-3}, 10^{-3})^T$
(7.1)	$\dot{q}_m = (0, 0, 0, 0, 0, 0)^T$
(7.39)	$q_s = (0.001, 0.001, 0, 0, 10^{-3}, 0)^T$
(7.39)	$\ddot{q}^1_m(0) = (0, 0, 0, 0, 0, 0)^T$
(7.66), (7.67)	$\tau_{1s}(0) = 0, \tau_{2s}(0) = 0$

Remark 7.5. Generally, the signum function is used as the corrective control but due its disadvantages 'tanh' is used instead. The 'tanh' function may also cause discontinuity due to its presence in the derivative of control signals (7.66) and (7.67). But, the actual control inputs are obtained after the integration and hence these control signals may not have the high frequency switching components (discontinuous term). Thus, the proposed SO-PID-TSMC technique ensures finite time stability with no chattering in the sliding surface (σ_i) and control input τ_{is}.

Remark 7.6. Here, the proposed sliding surfaces consist of proportional, derivative and integral terms of the conventional sliding surfaces. The integral part in the proposed sliding surface consists of the non singular term which helps to maintain the finite time convergence, the global stability of e_{is} and $\dot{e}_{is}$ along with low control energy consumption and zero offset error. The SO-PID-SMC does not have the singularity problem.

7.5 Results and discussion

The numerical simulation of the conventional sliding mode control and second order PID terminal sliding mode controllers have been performed using MATLAB®-15b. Various parametric uncertainties have been selected for testing the robustness of the controller. These uncertainties are considered in the range of $\pm 5\%$ to $\pm 35\%$ of their nominal values. It is observed that the controller can sustain the parametric uncertainty up to a range of $\pm 30\%$ as shown in the following figures. The proposed synchronisation strategy is validated on identical two-link flexible manipulators. The parameters of a planar TLFM used as both master and slave manipulators are given in Table 3.1. All the simulations are carried out in a MATLAB-14a environment using ode45 solver with a step size of 0.002. To validate the effectiveness of the proposed tip trajectory synchronisation, the desired tip trajectories are shown in Fig. 7.4. The desired tip trajectories are given in (7.83) and (7.84).

$$y_{d1} = \frac{\pi}{4} - \frac{7}{6}e^{-\frac{3}{2}t} + \frac{7}{19}e^{-2t} \tag{7.83}$$

$$y_{d2} = \frac{\pi}{6} - \frac{7}{5}e^{-\frac{3}{2}t} + \frac{7}{11}e^{-2t} \tag{7.84}$$

The initial conditions used for the simulation of the master, slave manipulators, and control inputs are given in Table 7.1. The various constant gains used in the control techniques and synchronisation are given as $c_{1m} = c_{2m} = 5$, $\rho_{1m} = \rho_{2m} = 10$, $c_{1s} = c_{2s} = 5$, $\rho_{11s} = \rho_{12s} = 10$, $\rho_{21s} = \rho_{22s} = 10$, $k_{D1} = k_{D2} = 1$, $k_{P1} = k_{P2} = 5$, $k_{I1} = k_{I2} = 5$, $\alpha_1 = \alpha_2 = 0.6$. These parameters are chosen such that the convergence rate (synchronisation time) and control energies are low. Tip trajectory tracking control for the master manipulator and synchronisation results are presented in the next subsections.

7.5.1 Tip trajectory tracking of the controlled master manipulator in the presence of parameters uncertainty

The tip trajectory tracking control of the master manipulator using conventional sliding mode control is presented in this section for a payload of 0.145 kg. The links of a flexible manipulator are made of steel. The density of steel is used to calculate the mass of the links. However, the density of the material may change when the composition of its alloys are changed, resulting in the change of the mass of the links. $\pm 5\%$ and $\pm 30\%$ uncertainties in the nominal mass values are taken under consideration. If the mass is changed, link moment of inertia and inertia of the link with respect to the joints will also change. Along with these parameters, the drive torque constant may also change due to its continuous use, wear and tear. The comparison of the tracking errors between the desired tip trajectory and the trajectories with nominal, $\pm 5\%$, $\pm 30\%$ parametric variations are given in Fig. 7.5.

It is seen that the steady state error is high in variations of -5% and -30% with 0.00002 rad as compared with almost zero in the case of nominal parameters. It also shows that the tip trajectory tracking control of the master manipulator with parametric uncertainties is successfully achieved within 1 s for both the links which is almost similar with the results of the nominal parameter. It is observed from Fig. 7.6, showing the results of tip deflections of the links of the master manipulator. The tip deflections of link-1 are completely suppressed within 5 s with a maximum deflection of ± 0.04 mm for the case with $\pm 30\%$ parametric uncertainties. In case of link-2, the tip deflections are in the range $+0.02$ mm to -0.025 mm for $\pm 30\%$ parametric uncertainties and are completely suppressed within 3 s. It is seen from Fig. 7.6 that the tip deflections are higher in cases of $+5\%$, $+30\%$ parametric uncertainties and lower in cases of -5%, -30% parametric uncertainties as compared with the nominal parametric uncertainties.

The sliding surfaces used for controlling the master manipulator for tracking the desired trajectories with $\pm 5\%$ and $\pm 30\%$ parametric uncertainties are shown in Fig. 7.7. It is seen that the reaching time for nominal and $+5\%$ parametric uncertainties is almost same with 0.25 s for link-1 and 0.2 s for link-2. The steady state error is in terms of 10^{-4} rad for link-1 and 10^{-3} rad for link-2. The reaching time is 1 s for link-1 and 0.8 s for link-2 with $+5\%$ uncertainty which is much slower than the nominal parameters. The comparisons of the torques are shown in Table 7.2.

Table 7.2: Comparison of the required control torques with and without parametric uncertainties for the master manipulator.

Link no.	without (Nm)	with $+5\%$ (Nm)	with -5% (Nm)	with $+30\%$ (Nm)	with -30% (Nm)
Link-1	13	13	13	14	40
Link-2	2.5	2.5	5	3.5	12

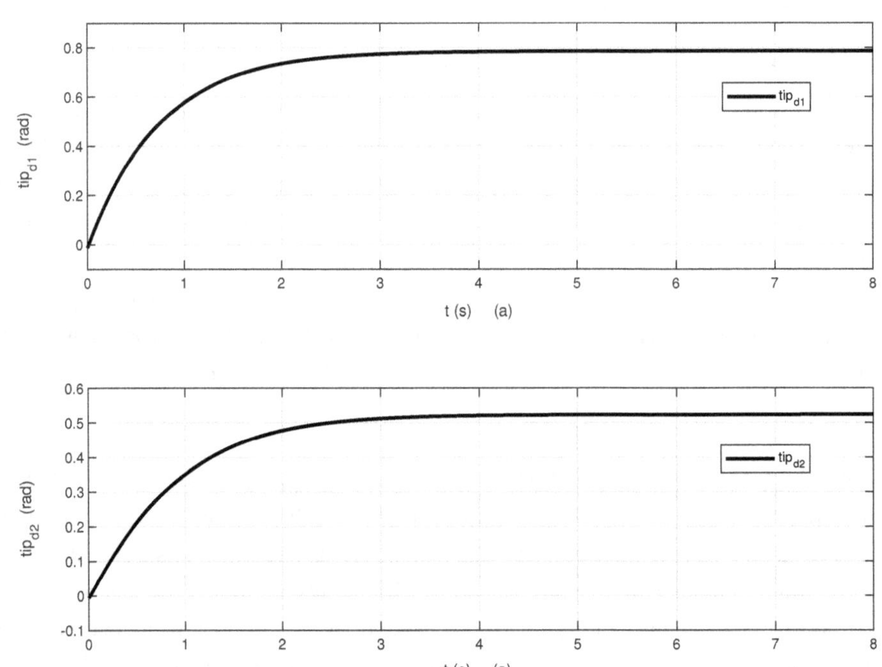

Figure 7.4: The desired tip trajectories of the master manipulator: (a) link-1 and (b) link-2.

7.5.2 Synchronisation of the controlled master (upto $+30\%$ parameters uncertainty) and slave (upto -30% parameters uncertainty) manipulators using SO-PID-TSMC

In this section, the synchronisation results between a controlled master and an identical slave TLFM are presented. There are two cases in this subsection. A payload of 0.145 kg is considered for the master and slave manipulator in both the cases.

Case-I is the synchronisation between a nominal controlled master and an identical slave. Case-II is the synchronisation between a controlled master with $+30\%$ parametric uncertainty and a slave with -30% parametric uncertainty. The comparative results of Case-I and Case-II are shown in subsequent figures. Figure 7.8 shows the comparison of the tip trajectory synchronisation error between Case-I and Case-II. It is apparent from the Fig. 7.8 that the tip trajectory synchronisation is achieved at 1.3 s for Case-I, at 3 s for Case-II for both the links, respectively. The tip deflections of the masters and slaves are given in Fig. 7.9. It is noted from Fig. 7.9 that the tip deflections

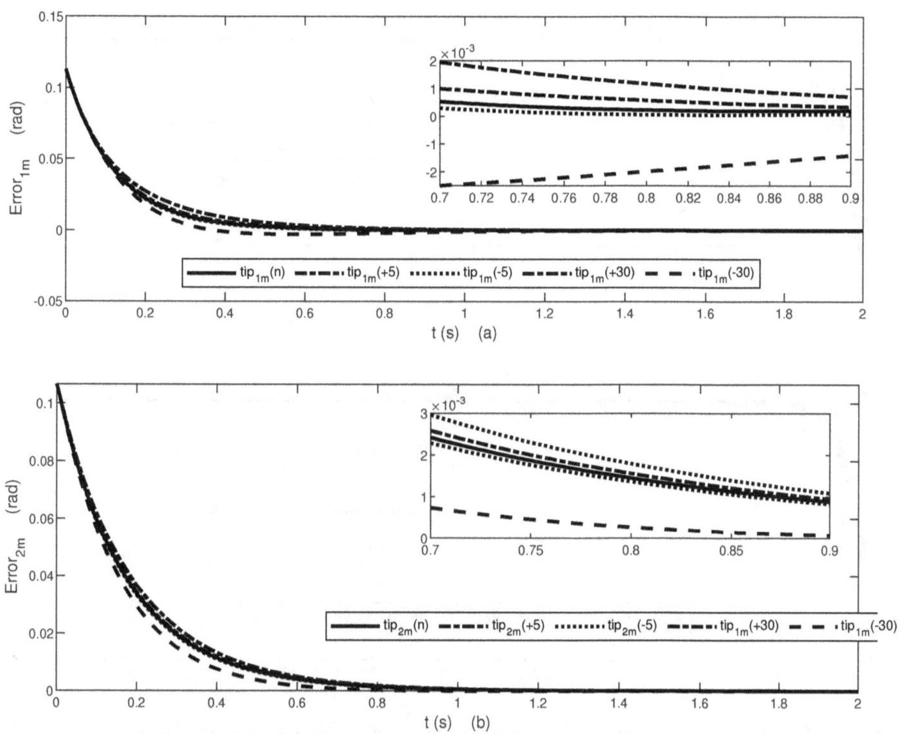

Figure 7.5: Comparison of the tracking errors between the desired tip trajectory and the trajectories of the master manipulator with nominal, ±5% and ±30% uncertainties in the parameters: (a) link-1 and (b) link-2.

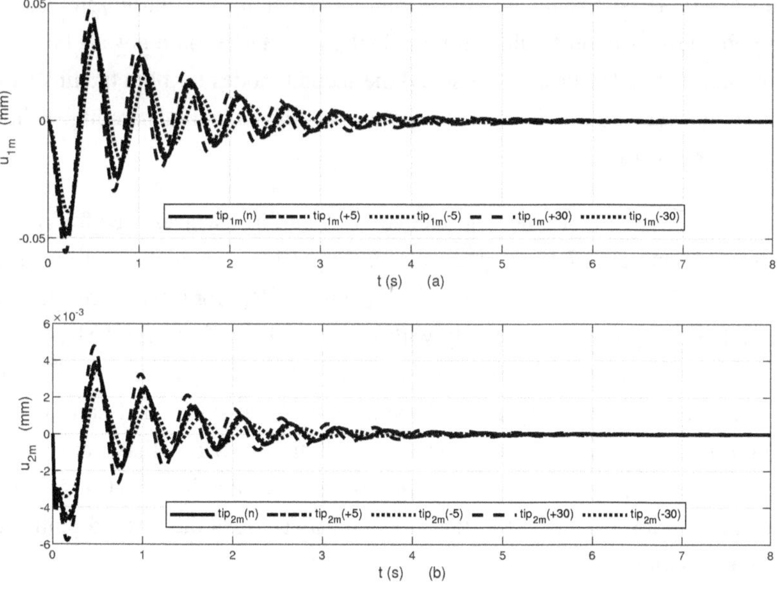

Figure 7.6: Tip deflections of the links of the master manipulator having 0.145 kg payload with nominal, ±5% and ±30% uncertainties in the parameters: (a) link-1 and (b) link-2.

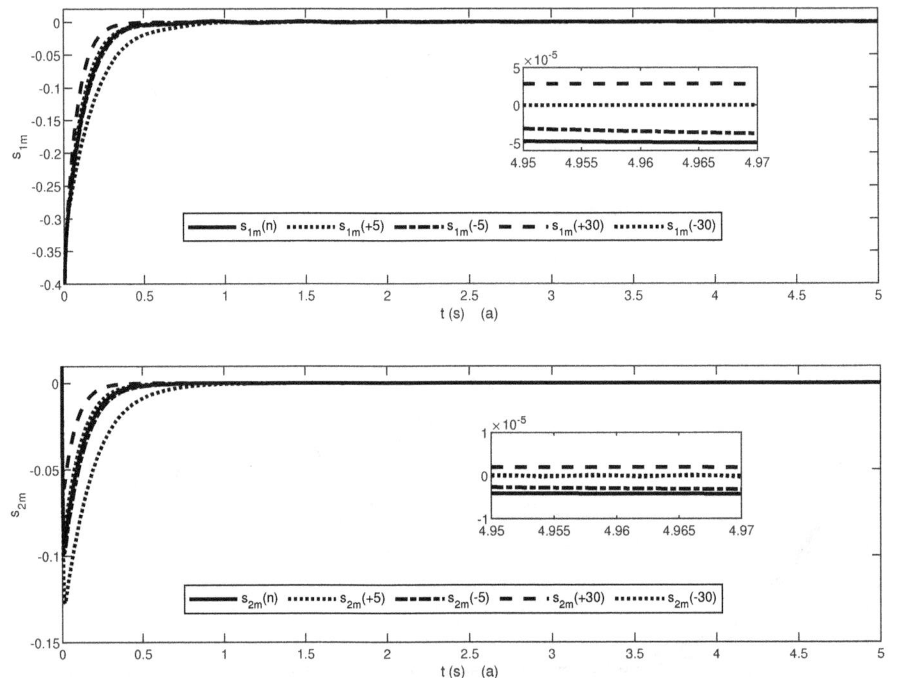

Figure 7.7: Sliding surfaces of the master manipulator having 0.145 kg payload with nominal ±5% and ±30% uncertainties in the parameter: (a) link-1 and (b) link-2.

are suppressed in the range of ±0.06 mm for link-1 and [−0.012 to +0.006] mm for link-2 of the manipulator in Case-II. These results are higher than the deflections of master and slave in Case-I. Such results are obvious since the results with nominal parameters should always be better than the results with parametric uncertainties. The first and the second modes for both the links are shown in Fig. 7.10 for both the Cases. In Fig. 7.10, it is apparent that the deflections are higher in the case of the slave manipulator of Case-II.

The responses of the proposed first order and second order sliding surfaces used for synchronisation are shown in Fig. 7.11 and Fig. 7.12, respectively. In Fig. 7.11, it is shown that the reaching time for Case-I is 2.7 s and 2.5 s for link-1 and link-2, respectively. But for Case-II, reaching time is 3.8 s and 3.6 s for link-1 and link-2, respectively with no chattering effect. Similarly, from Fig. 7.12, it is clear that the reaching time for the designed second order sliding surface for Case-I is 3 s and 3.2 s for link-1 and link-2, respectively, and 3.2 s and 4 s for Case-II of link-1 and link-2. Although there are variations in the results of Case-I and Case-II, all these results satisfy our objectives. From the above discussed results, it reveals that the proposed controller is robust and can handle the parametric uncertainties up to ±30%. Further, the controller tracks the desired path satisfactorily with proper synchronisation.

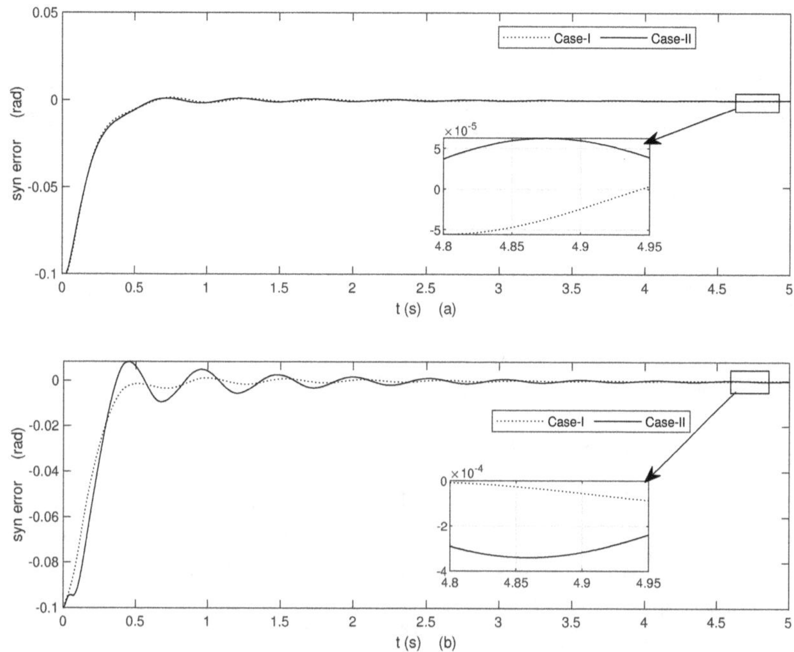

Figure 7.8: Comparison of the tip trajectory synchronisation errors between Case-I and Case-II: (a) link-1 and (b) link-2.

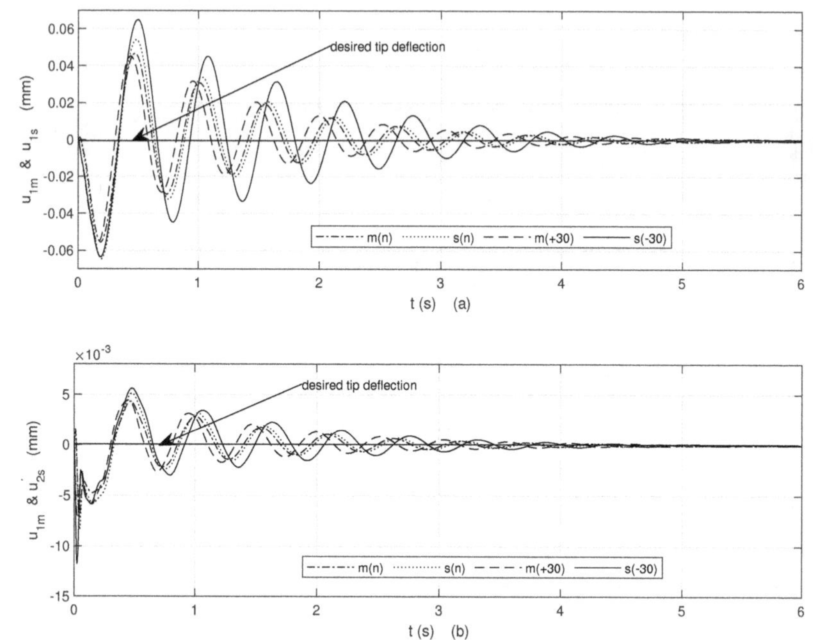

Figure 7.9: Tip deflections of the links in Case-I and Case-II: (a) for link-1 and (b) link-2.

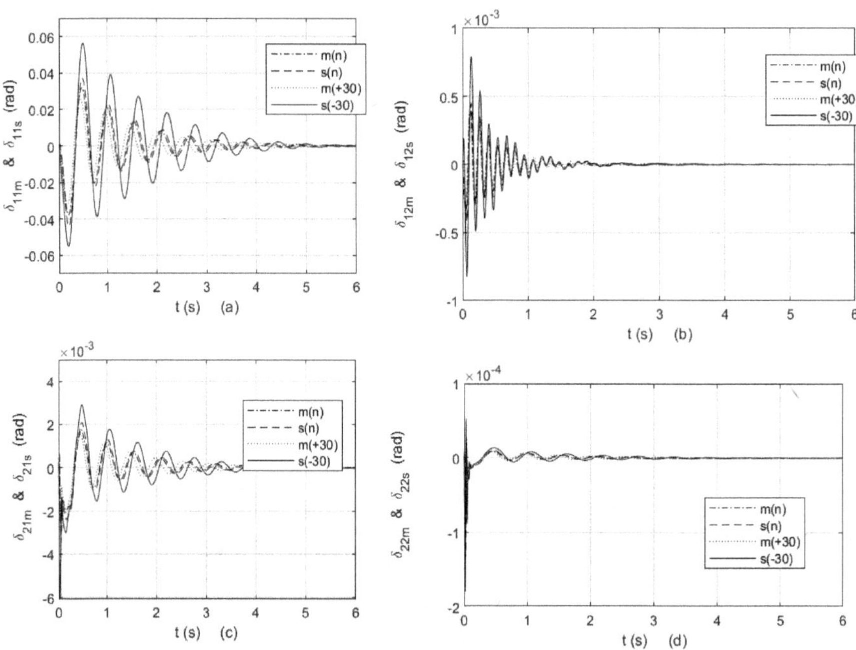

Figure 7.10: Modes of the links in Case-I and Case-II: (a) mode-1 of link-1, (b) mode-2 of link-1, (c) mode-1 of link-2 and (d) mode-2 of link-2.

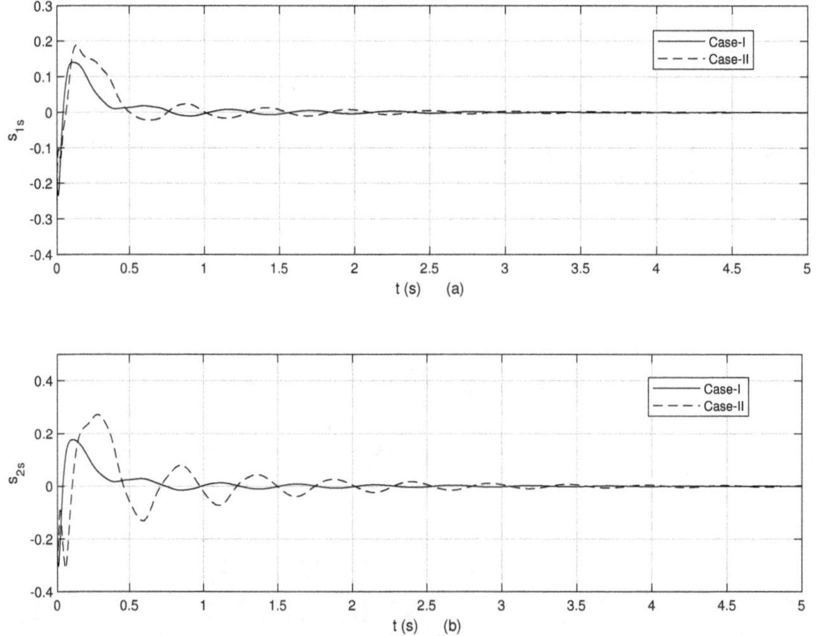

Figure 7.11: The first order sliding surfaces during synchronisation of Case-I and Case-II: (a) link-1 and (b) link-2.

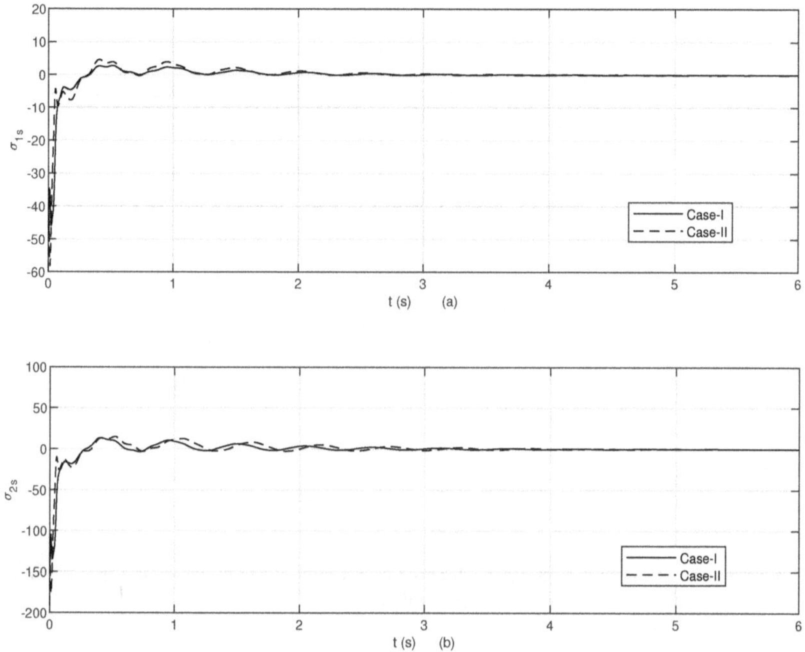

Figure 7.12: The second order sliding surfaces during synchronisation of Case-I and Case-II: (a) link-1 and (b) link-2.

7.5.3 Synchronisation of the controlled master TLFM having 0.145 kg payload and the slave having 0.3 kg payload TLFM

The synchronisation results between the controlled master TLFM having 0.145 kg payload and a slave TLFM having 0.3 kg payload are presented in this subsection. Further, for comparison of robustness of the proposed controller, two cases are considered here. Case-A is the synchronisation between the nominal controlled master and slave manipulators with a 0.145 kg payload each. Case-B is the synchronisation between a controlled master manipulator with a 0.145 kg payload and a slave manipulator with a 0.3 kg payload. The comparative results of Case-A and Case-B are shown in the subsequent figures.

Figure 7.13 shows the comparison of synchronisation errors between cases A and B. The synchronisation is obtained at 1.2 s for both the Cases with a steady state error of the order of 10^{-4} rad for link-1. Similarly, for link-2, the synchronisation is achieved at 2 s for both the Cases with a steady state error of the order of 10^{-4} rad. The tip deflections of the controlled master and slave during synchronisation for both the Cases are shown in Fig. 7.14. As there is an additional 0.155 kg payload at the slave manipulator in Case-B, the deflections should be more in it which is evident from the figure. The deflections of the slave manipulator range from -0.09 mm to $+0.08$ mm for link-1 and -0.013 mm to 0.006 mm for link-2 in Case-B. The settling time is 7.8 s for link-1 and 6.5 s for link-2. The settling time is almost the same in link-1 and link-2 for cases A and B. There is a difference of 0.01 mm in the case of the master manipulator for both in cases A and B and the slave

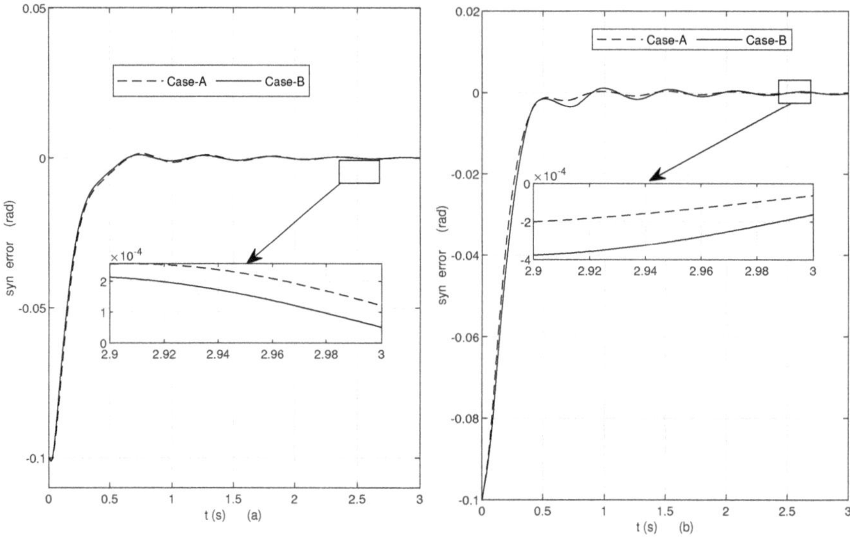

Figure 7.13: Comparison of the tip trajectory synchronisation errors between Case-A and Case-B: (a) link-1 and (b) link-2.

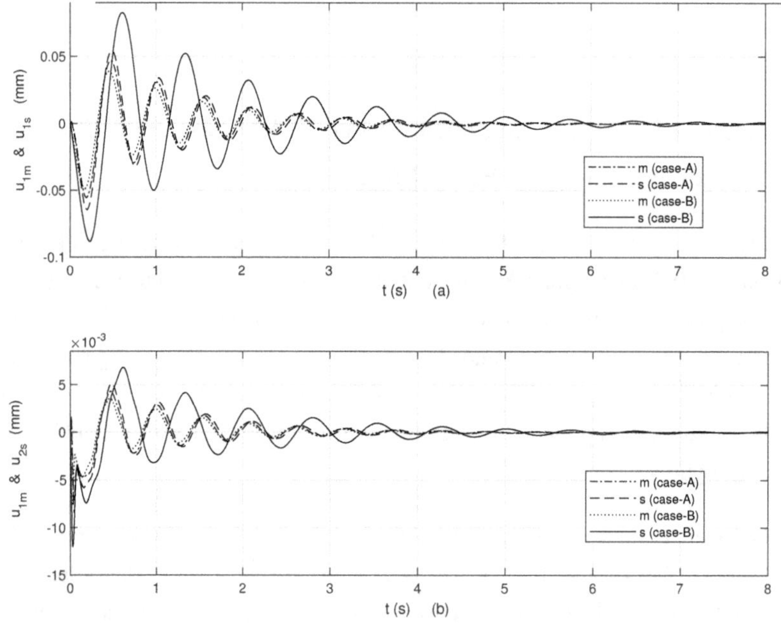

Figure 7.14: Comparison of the tip deflections of the links in Case-A and Case-B: (a) link-1 and (b) link-2.

manipulator for Case-A in terms of tip deflections. Figure 7.15 shows the results of the responses of the proposed second order sliding surfaces for the synchronisation. It is seen from Fig. 7.15 that the settling time as well as the amplitude of the second order sliding surfaces are more in Case-B as compared with case-A for both the links.

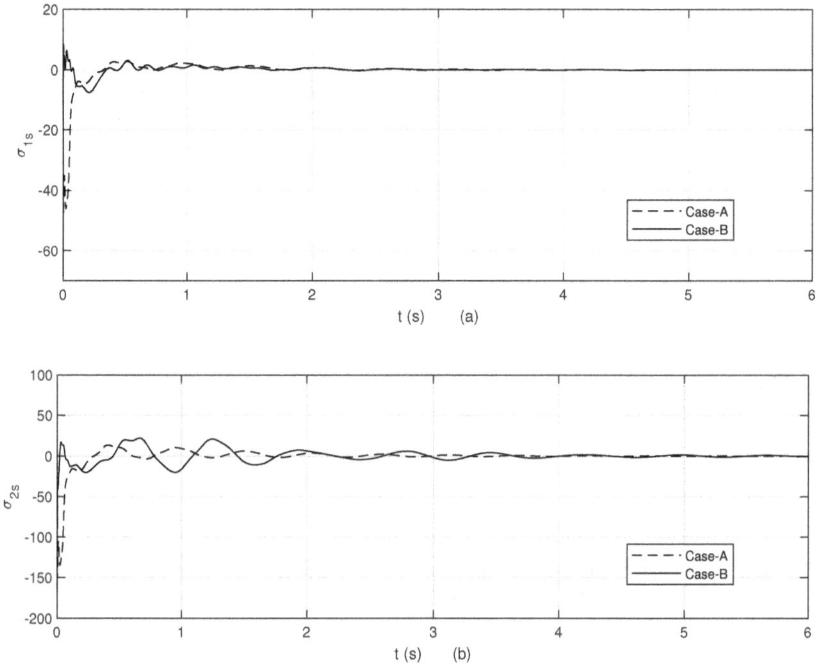

Figure 7.15: Comparison of the second order sliding surfaces between Case-A and Case-B: (a) link-1 and (b) link-2.

7.5.4 Comparison between the proposed SO-PID-TSMC and the SO-SMC

In this section, a comparison between the proposed SO-PID-TSMC and the normal SO-SMC is presented. The master manipulator is given a 0.145 kg payload and the slave manipulator is given a 0.145 kg payload. The structure of the normal second order sliding surfaces used for comparison is considered as:

$$\sigma_{1o} = \dot{s}_{1s} + k_1 s_{1s} \tag{7.85}$$

$$\sigma_{2o} = \dot{s}_{2s} + k_2 s_{2s} \tag{7.86}$$

where $k_1 = 5, k_2 = 5$ are chosen here as the design parameters. The tip synchronisation errors between the controlled master and slave manipulators using both the controllers are shown in Fig. 7.16. Figure 7.17 shows the tip deflections for the synchronised slave manipulator using both the controllers. The second order sliding surfaces and required torque inputs for synchronisation using both the controllers are shown in Fig. 7.18 and Fig. 7.19, respectively. For comparing the performances of the controllers, three performance measures are considered. These are (i) integral time absolute error (ITAE) of synchronisation, (ii) total variation (TV) in the control input and (iii) two norm of the control input (u2 norm). The performances are calculated for 10 time units with 0.001 sample size. The results of the performances of the controllers are given in Table 7.3 for comparison. It is apparent from the Table 7.3 that the performances of the proposed SO-PID-TSM controller are better than the existing normal second order SMC.

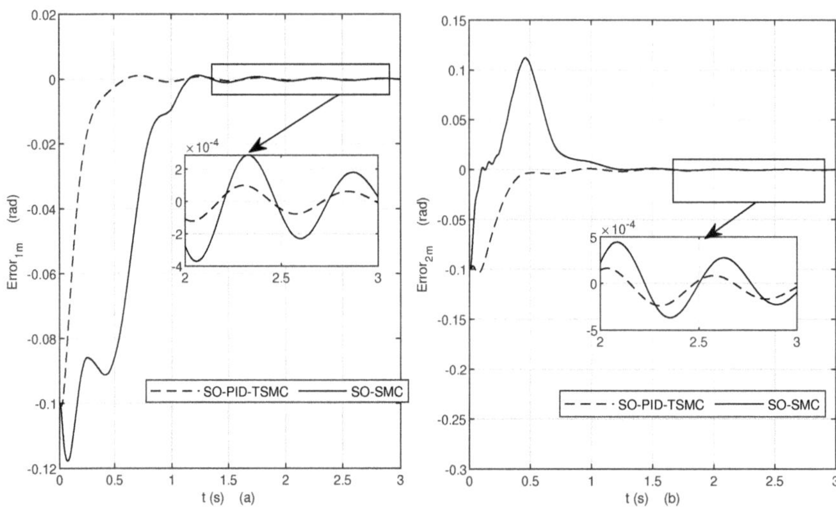

Figure 7.16: Tip trajectory synchronisation errors between the nominal controlled master and the nominal slave manipulators using SO-PID-TSMC and SO-SMC.

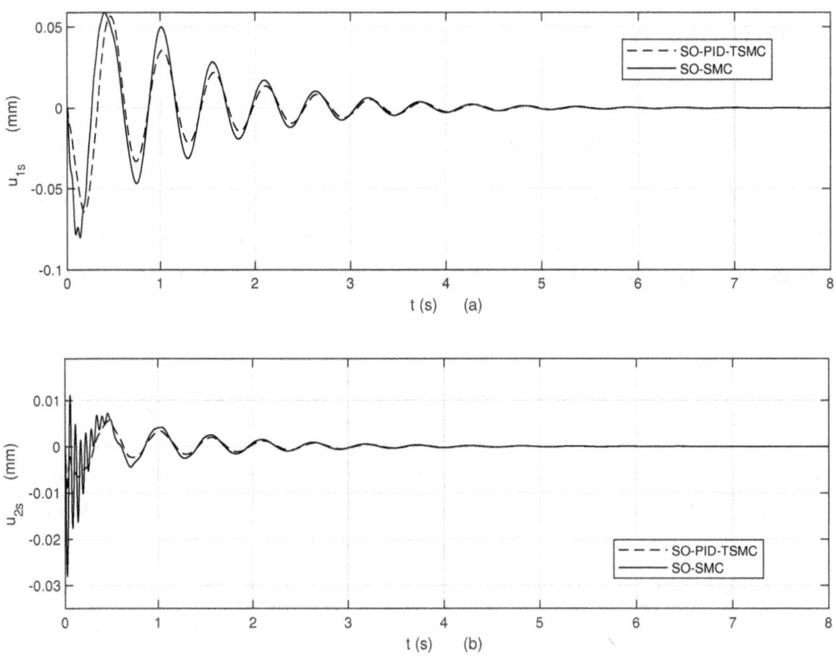

Figure 7.17: Comparison between the tip deflections of links of the nominal slave manipulators using SO-PID-TSMC and SO-SMC.

It is seen from Fig. 7.16 and Table 7.3 that the synchronisation is achieved early using the SO-PID-TSMC in comparison with SO-SMC. The synchronisation error is also less in SO-PID-TSMC. Further, the control efforts for the proposed control technique is less in comparison with that of

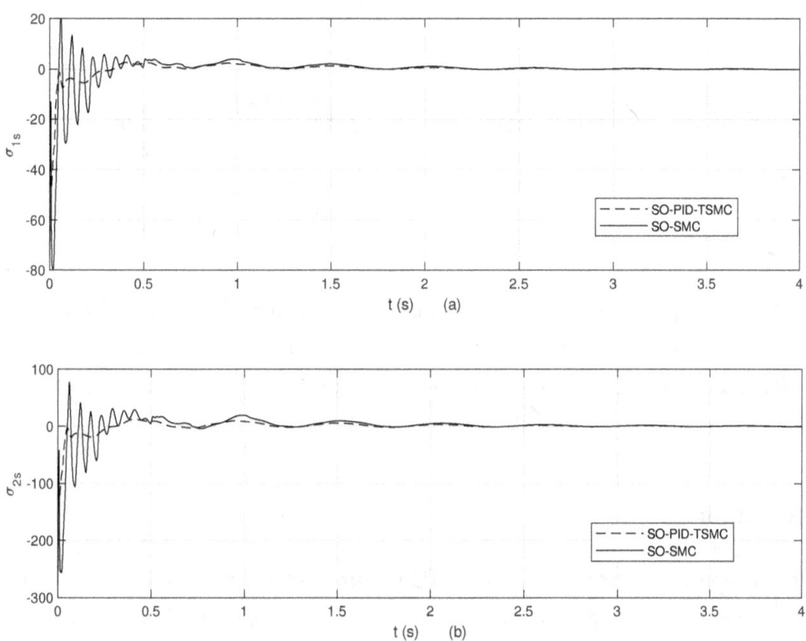

Figure 7.18: Comparison between the second order sliding surfaces using SO-PID-TSMC and SO-SMC.

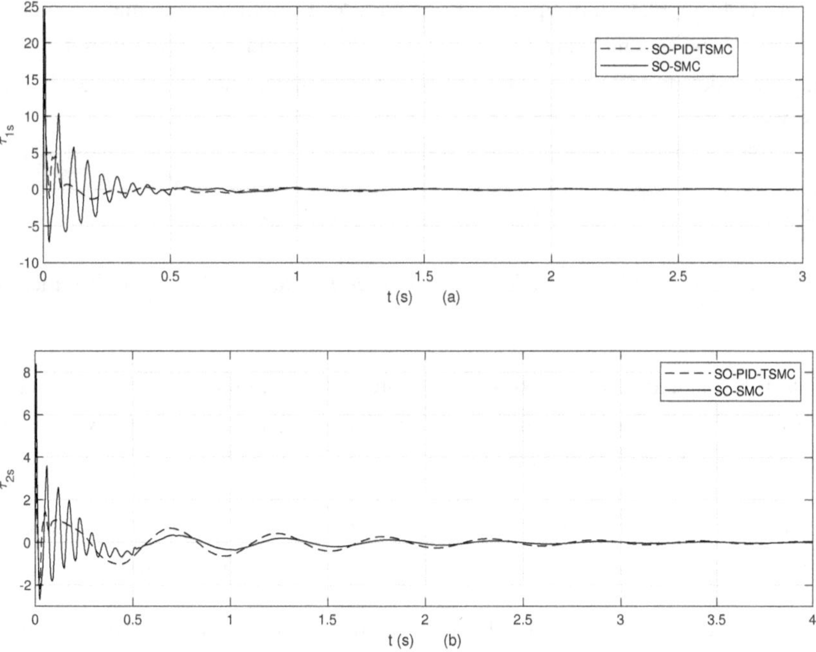

Figure 7.19: Comparison between the required torque inputs for synchronisation using SO-PID-TSMC and SO-SMC.

Table 7.3: Comparison of performances between the controllers.

Control technique	link	ITAE of synchronisation	Total variation (TV) in control signal	Torque energy
SO-PID-TSMC	link-1	**0.0035**	**56.794**	**49.385**
	link-2	**0.007**	**29.260**	**25.777**
SO-SMC	link-1	0.025	133.692	118.503
	link-2	0.021	49.681	44.917

the SO-SMC. Thus, the proposed SO-PID-TSM controller achieved good synchronisation performances by using less and smoother control inputs. Since SO-SMC is considered as a very good robust controller than other controllers reported in the literature, the comparison of the proposed SO-PID-SMC is restricted to the SO-SMC controller only.

7.6 Chapter summary

This chapter has proposed a new second order PID terminal SMC for tip trajectory synchronisation along with the suppression of tip deflection between two identical two-link flexible manipulators with variable payloads and model parametric uncertainties. The proposed sliding surfaces of the SO-PID-SMC are designed in terms of normal sliding surfaces. The simulation results show that the proposed control technique has the ability for a finite time convergence of the synchronisation error along with fast suppression of tip deflection and global stability of the manipulator dynamics. The performance of the second order PID terminal SMC is evaluated in the presence of bounded parameter uncertainties and variable payloads. The performance of SO-PID-SMC is found to be better than SO-SMC.

References

[1] M. O. Tokhi and A. K. M. Azad. *Flexible robot manipulators*. IET, Control Engineering Series 68, 2002.

[2] A. Rodriguez-Angeles and H. Nijmeijer. Mutual synchronization of robots via estimated state feedback: A cooperative approach. *IEEE Transactions on Control Systems Technology*, vol. 12, no. 4, pp. 542–554, 2004.

[3] B. Subudhi and A. Morris. Soft computing methods applied to the control of a flexible robot manipulator. *Applied Soft Computing*, vol. 9, no. 1, pp. 149–158, 2009.

[4] S. K. Pradhan and B. Subudhi. Real-Time adaptive control of a flexible manipulator using reinforcement learning. *IEEE Transactions on Automation Science and Engineering*, vol. 9, no. 2, pp. 237–249, 2012.

[5] W. Shang, S. Cong, and Y. Ge. Coordination motion control in the task space for parallel manipulators with actuation redundancy. *IEEE Transactions on Automation Science and Engineering*, vol. 10, no. 3, pp. 665–673, 2013.

[6] D. Zhao and Q. Zhu. Position synchronised control of multiple robotic manipulators based on integral sliding mode. *International Journal of Systems Science*, vol. 45, no. January 2014, pp. 556–570, 2014.

[7] L. Chao, Z. Dongya, and X. Xianbo. Force synchronization of multiple robot manipulators: A first study. *Proceedings of the 33rd Chinese Control Conference*, pp. 2212–2217, 2014.

[8] D. Zhao, S. Li, and Q. Zhu. Adaptive synchronised tracking control for multiple robotic manipulators with uncertain kinematics and dynamics. *International Journal of Systems Science*, vol. 7721, no. May 2014, pp. 1–14, 2014.

[9] H. Dou and S. Wang. Robust adaptive motion/force control for motion synchronization of multiple uncertain two-link manipulators. *Mechanism and Machine Theory*, vol. 67, pp. 77–93, 2013.

[10] H. Dou and S. Wang. A boundary control for motion synchronization of a two-manipulator system with a flexible beam. *Automatica*, vol. 50, no. 12, pp. 3088–3099, 2014.

[11] Q. Cao, S. Li, D. Zhao, and Z. Wang. Finite-time motion/force control for motion synchronization of multiple manipulators. *Proceedings of the 33rd Chinese Control Conference, CCC 2014*, pp. 2121–2126, 2014.

[12] R. Cui and W. Yan. Mutual synchronization of multiple robot manipulators with unknown dynamics. *Journal of Intelligent and Robotic Systems: Theory and Applications*, vol. 68, no. 2, pp. 105–119, 2012.

[13] L. D. Khoa, D. Q. Truong, and K. K. Ahn. Synchronization controller for a 3-R planar parallel pneumatic artificial muscle (PAM) robot using modified ANFIS algorithm. *Mechatronics*, vol. 23, no. 4, pp. 462–479, 2013.

[14] D. Zhao, S. Li, and Q. Zhu. A new TSMC prototype robust nonlinear task space control of a 6 DOF parallel robotic manipulator. *International Journal of Control, Automation and Systems*, vol. 8, no. 6, pp. 1189–1197, 2010.

[15] T. Endo, F. Matsuno, and H. Kawasaki. Simple boundary cooperative control of two one-link flexible arms for grasping. *IEEE Transactions on Automatic Control*, vol. 54, no. 10, pp. 2470–2476, 2009.

[16] Y. Su, D. Sun, L. Ren, and J. K. Mills. Integration of saturated PI synchronous control and PD feedback for control of parallel manipulators. *IEEE Transactions on Robotics*, vol. 22, no. 1, pp. 202–207, 2006.

[17] W. Shang, S. Cong, Y. Zhang, and Y. Liang. Active joint synchronization control for a 2-DOF redundantly actuated parallel manipulator. *IEEE Transactions on Control Systems Technology*, vol. 17, no. 2, pp. 416–423, 2009.

[18] X. T. Tran and H. J. Kang. Adaptive hybrid high-Order terminal sliding mode control of MIMO uncertain nonlinear systems and its application to robot manipulators. *International Journal of Precision Engineering and Manufacturing*, vol. 16, no. 2, pp. 255–266, 2015.

[19] S. Ding, J. Wang, and W. X. Zheng. Second-Order sliding mode control for nonlinear uncertain systems bounded by positive functions. *IEEE Transactions on Industrial Electronics*, vol. 62, no. 9, pp. 5899–5909, 2015.

[20] A. Nasiri, S. Kiong Nguang, and A. Swain. Adaptive sliding mode control for a class of MIMO nonlinear systems with uncertainties. *Journal of the Franklin Institute*, vol. 351, no. 4, pp. 2048–2061, 2014.

[21] H. Alwi and C. Edwards. Sliding mode fault-tolerant control of an octorotor using linear parameter varying-based schemes. *IET Control Theory & Applications*, vol. 9, no. 4, pp. 618–636, 2015.

[22] D. Zhao, T. Zou, S. Li, and Q. Zhu. Adaptive backstepping sliding mode control for leader–follower multi-agent systems. *IET Control Theory & Applications*, vol. 6, no. January 2011, p. 1109, 2012.

[23] D. Zhao, C. Li, and Q. Zhu. Low-pass-filter-based position synchronization sliding mode control for multiple robotic manipulator systems. *Proceedings of the Institution of Mechanical Engineers, Part I: Journal of Systems and Control Engineering*, vol. 225, no. 8, pp. 1136–1148, 2011.

[24] R. Su and Q. Zong. Comprehensive design of disturbance observer and non-singular terminal sliding. *IET Control Theory & Applications*, vol. 9, no. 12, pp. 1821–1830, 2015.

[25] A. Al-Ghanimi, Z. Man, and J. Zheng. Robust and fast non-singular terminal sliding mode control for piezoelectric actuators. *IET Control Theory & Applications*, vol. 9, no. 18, pp. 2678–2687, 2015.

[26] Y. Wang, C. Wang, P. Lu, and Y. Wang. High-order nonsingular terminal sliding mode optimal control of two-link flexible manipulators. *IECON Proceedings (Industrial Electronics Conference)*, pp. 3953–3958, 2011.

[27] Y. Wang, H. Xia, and C. Wang. Communications in computer and information science. In *Applied Informatics and Communication*, pp. 409–418, 2011.

[28] Y. Wang, Y. Cao, and H. Xia. Optimized continuous non-singular terminal sliding mode control of uncertain flexible manipulators. *Chinese Control Conference, CCC*, vol. 2015-September, pp. 3392–3397, 2015.

[29] Y. Wang, F. Han, Y. Feng, and H. Xia. Hybrid continuous nonsingular terminal sliding mode control of uncertain flexible manipulators. *Proceedings, IECON 2014 - 40th Annual Conference of the IEEE Industrial Electronics Society*, no. 51307035, pp. 190–196, 2014.

[30] S. Rui, Z. Qun, T. Bailing, and Y. Ming. Comprehensive design of disturbance observer and nonsingular terminal sliding. *IET Control Theory Appl.*, vol. 9, no. 12, pp. 1821–1830, 2015.

[31] A. Ali, Z. Jinchuan, and M. Zhihong. Robust and fast non-singular terminal sliding mode control for piezoelectric actuators. *IET Control Theory Appl.*, vol. 9, no. 18, pp. 2678–2687, 2015.

[32] H. K. Khalil. *Nonlinear Systems*, Third ed., Prentice Hall, Upper Saddle River, New Jersey, 2002.

Chapter 8

Projective Synchronisation between Assumed Modes and Lumped Parameter Modelled TLFMs

It is seen from Chapter 6 and 7 that the performance and the desired operation of the synchronised manipulator depends on the dynamical analyses of its corresponding mathematical model. Improved modelling causes degradation in the response of the system. It is noted from Chapter 4 that modelling of a flexible manipulator is also the area of focused research in the field of manipulators. It is also noted from Chapters 6 and 7 that for purpose of synchronisation the dynamic structures of the both the master and slave manipulators are considered as same. But, the problem arises when the dynamic structure of the master and slave manipulators are different. In such a scenario, synchronising two different manipulators is a very challenging task. In this chapter, such a case is considered, where a master manipulator is modelled using an AMM having six states and slave manipulators are modelled using the LPM method having four states.

The chapter is organised as the chapter objective, the design of an adaptive SMC based filter for tracking control of master manipulator. The design of an adaptive global super-twisting sliding mode based filter for the synchronisation of controlled master and slave manipulator is also discussed. The simulations are done for the designed controllers and the results and discussion are given and finally, the chapter summary is presented at the end.

8.1 Introduction

Robustness and the precise operation of the system depend on the designed control algorithm. The synchronisation of the robot manipulators is considered a bilateral teleportation system [1]. In this case, a master manipulator is operated by a human and provides motion command to the slave manipulator [2]. The synchronisation between the master and slave manipulators is considered as the main control objective in teleportation systems [3]. During synchronisation and control, many uncertainties and complexities occur.

From the last two chapters it is seen that the operation of any synchronisation strategy depends on the type of control algorithm used [4]. It is also seen that many control techniques are designed for the synchronisation of a rigid robot manipulator like adaptive control [5, 6], observer based control and feedback control [7], feedback with PD and saturated PI [8] and computed torque control [9]. As discussed in Chapters 6 and 7 several SMC techniques are used for synchronisation of rigid robot manipulators like SMC with adaptive control in [10–14], adaptive backstepping SMC in [15], low pass filter based SMC in [16], integral SMC in [17], SMC with fuzzy logic and H_∞ in [18]. A conventional SMC has ineffective performances in the reaching phase of the sliding surface and may destabilise the controlled system [19]. In such a case, the global sliding mode (GSM) control scheme is considered the best, because it eliminates the reaching mode as the sliding mode occurs from a starting point. Thus, the system response is insensitive to uncertainties and disturbances [20]. Recently, many works are reported on the GSMC for different control problems [21].

The reported works on the synchronisation of the manipulators are categorised in the Table 6.1.

Motivation:

- From Table 6.1, we see that the synchronisation between the flexible manipulators (FMs) is a new approach for the industrial application of the flexible manipulator. The projective synchronisation where twelve states are projected into eight states is also not seen in the literature where the master and slave manipulators are modelled differently.

- It is also seen that adaptive time-varying super-twisting GSMC has not yet been attempted in the literature of flexible manipulators. The fact that the gains of the sliding surfaces and corrective control law are updated online which stabilises the system within a fixed time motivated us to design an adaptive time-varying super-twisting GSMC.

8.2 Chapter objectives

To address the synchronisation control problem outlined in this chapter, the following objectives have been defined:

- To design a global sliding mode controller for the master manipulator to track the desired trajectory.

- To design an adaptive time-varying super-twisting global sliding mode controller for the synchronisation of the controlled master and the differently modelled slave manipulators (two identical and one non-identical to the master manipulator).
- To compare the performances of the designed controllers with an existing controller [20].

8.3 Tracking control of a master manipulator using global SMC

The block diagram of the complete synchronisation strategy is shown in Fig 8.1. The dynamic equations of the master manipulator (7.1), can be written as

$$\begin{aligned}\ddot{g}_m &= B_{0m}^{-1}(g_m)[-h_{0m}(g_m,\dot{g}_m) - K_{0m}g_m - D_{0m}\dot{g}_m \\ &\quad + b\tau_m + b\tau_{dm} + Q_m(g,\dot{g},\ddot{g})]\end{aligned} \tag{8.1}$$

where $g_m = [\phi_{1m},\phi_{2m},z_{11m},z_{12m},z_{21m},z_{22m}]^T$, and τ_m is the required control input torque for the master manipulator. The elastic deflection of the master manipulator (8.1) can be written as

$$\xi_{im} = \sum_{j=1}^{2} w_{ijm}z_{ijm} \tag{8.2}$$

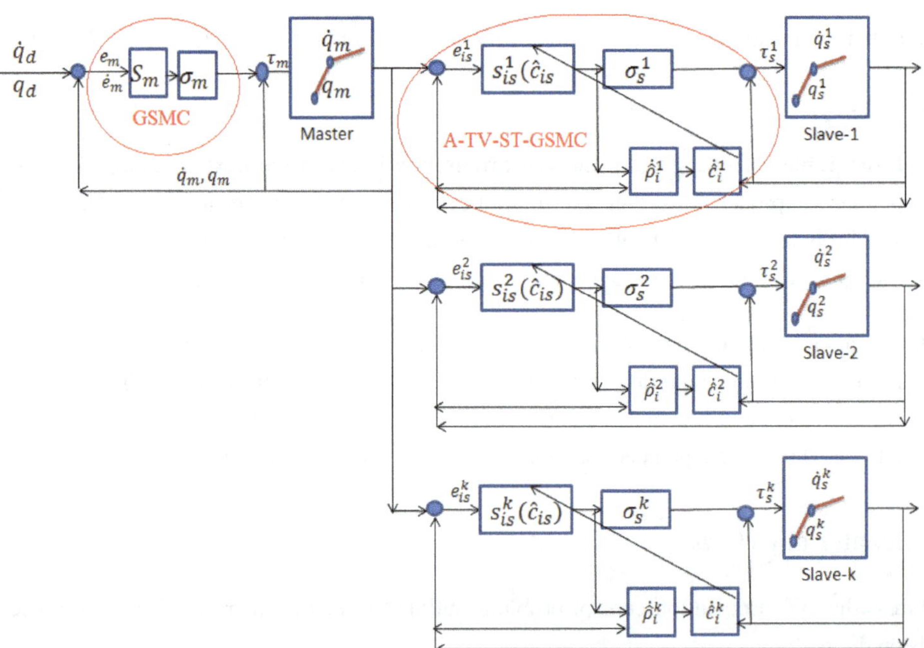

Figure 8.1: The block diagram of the complete synchronisation strategy between the controlled master and the slave manipulators.

Let ϕ_{di} be the desired trajectories of the links and its expression is given in the results and discussion section. The tracking errors are,

$$e_{im} = \phi_{im} - \phi_{di} \tag{8.3}$$

Here, the aim is to determine τ_m by designing GSMC for the convergence of error dynamics (8.3) to zero in the required time duration.

In the first step of designing GSMC, normal SMC sliding surfaces are developed as follows

$$s_{im} = \dot{e}_{im} + c_{im}e_{im} \tag{8.4}$$

where $c_{im} > 0$ are user defined constant gains given in the results and discussion section. For the existence of sliding mode, the necessary and sufficient condition is to satisfy $s_{im} = \dot{s}_{im} = 0$. When (8.1) reaches the sliding surface, the equivalent sliding mode dynamics can be written as

$$\ddot{e}_{im} = -c_{im}\dot{e}_{im} \tag{8.5}$$

Remark 8.1. The stability of the equivalent sliding mode dynamics (8.5) can be defined by considering a Lyapunov function candidate whose time derivative is given as $\dot{V}_1 = -c_{im}\dot{e}_{im}^2$. Thus, the sliding mode dynamics (8.4) are asymptotically stable and the trajectories converge to the equilibrium point, $e_{1m} = e_{2m} = 0$.

Using (8.4), sliding surface of GSMC is designed as in (8.6). Now, the global sliding mode control (GSMC) is designed as (8.6),

$$\lambda_i(t) = \chi_i(s_i(t) - s_i(0)e^{-\mu_i t}) \tag{8.6}$$

The equivalent sliding mode dynamics of (8.6) is defined as

$$\chi_i(s_i(t) - s_i(0)e^{-\mu_i t}) = 0 \tag{8.7}$$

The constants χ_i and μ_i are the positive gain constants and satisfy the Hurwitz polynomial

$$X(\varphi) = \chi_{i_{n-2}}\Upsilon^{n-2} + \ldots + \chi_{i_1}\Psi + \chi_{io} \tag{8.8}$$

Theorem 8.2. *The tracking errors in (8.3) are asymptotically stable within the reaching time, if the sliding surfaces are chosen as in (8.6) with the control law as given in (8.9) and (8.10).*

$$\tau_1 = \frac{A_2N_{22} - A_4N_{12}}{N_{11}N_{22} - N_{21}N_{12}} - \rho_{1m}tanh(\lambda_{1m}) \tag{8.9}$$

$$\tau_2 = \frac{A_2}{N_{12}} - \frac{N_{11}}{N_{12}}(\frac{A_2N_{22} - A_4N_{12}}{N_{11}N_{22} - N_{21}N_{12}}) - \rho_{2m}tanh(\lambda_{2m}) \tag{8.10}$$

where

$$\begin{cases} qA = N_m(Hq + Kq + D\dot{q}) \\ A_1 = \ddot{\theta}_d - c_1\dot{e}_1 - \mu_{1m}e^{\mu_{1m}t}s_{1m}(0) \\ A_2 = qA(1) + A_1 \\ A_3 = \ddot{\theta}_d - c_2\dot{e}_2 - \mu_{2m}e^{\mu_{2m}t}s_{2m}(0) \\ A_4 = qA(2) + A_3 \end{cases}$$

where $N_{ij} = B_0^{-1}$, $qA(i)$ *are the elements of* qA.

Proof. The derivative of the sliding surface λ_i is given as,

$$\dot{\lambda}_i = \chi_i(\dot{s}_i(t) + s_i(0)\mu_i e^{-\mu_i t}) \tag{8.11}$$

Now, using the derivative of (8.4) along with (8.3), we can write it as

$$\dot{\lambda}_i = \chi_i(\ddot{e}_i + c_i\dot{e}_i + \mu_i e^{-\mu_i t}s_i(0)) \tag{8.12}$$

since, $\ddot{e}_i = \ddot{\phi}_i - \ddot{\phi}_{di} = \ddot{g}_m(i) - \ddot{\phi}_{di}$. Considering, the Lyapunov function candidate

$$V_2 = \frac{1}{2}\lambda_i\lambda_i^T \tag{8.13}$$

Taking the time derivative of (8.13) and using (8.9), (8.10), we get,

$$\dot{V}_2 = \lambda_i(-Q_m(g, \dot{g}, \ddot{g}) - \rho_{im}tanh(\lambda_{im})) \tag{8.14}$$

From (3.49), we get,

$$\gamma_0 + \gamma_1\|q\| + \gamma_2\|\dot{q}\|^2 = R \tag{8.15}$$

Then,

$$\begin{aligned} \dot{V}_2 &= -R|\lambda_{im}| - \lambda_{im}\rho_{im}tanh(\lambda_{im}) \\ &= -R|\lambda_{im}| - \rho_{im}|\lambda_{im}| \\ &= -|\lambda_{im}|(R + \rho_{im}) \\ &= -|\lambda_{im}|\Gamma \end{aligned} \tag{8.16}$$

where $\Gamma = R + \rho_{im}$. Thus, according to the Lyapunov stability theory, (8.16) is a negative definite function. Hence, the master manipulators (8.1) follow the desired trajectory. □

8.4 Projective synchronisation of the controlled master and three slave manipulators

This section describes the synchronisation between one controlled master and the three slaves (two identical to the master and one non-identical). The dynamics of the slave manipulators are written as in (8.17)

$$\ddot{g}_s^k = \begin{bmatrix} -b_1^k & b_2^k & 0 & 0 \\ 0 & 0 & -\ b_4^k & b_5^k \\ b_1^k & -b_7^k & 0 & 0 \\ 0 & 0 & b_4^k & -b_8^k \end{bmatrix} \begin{bmatrix} \dot{\phi}_{1s}^k \\ \xi_{1s}^k \\ \dot{\phi}_{1s}^k \\ \xi_{2s}^k \end{bmatrix} + d_{is}^k + \begin{bmatrix} b_3^k \\ b_6^k \\ -b_3^k \\ -b_6^k \end{bmatrix} u_s^k \tag{8.17}$$

where $g_s^k = [\phi_{1s}^k, \phi_{2s}^k, \xi_{1s}^k, \xi_{2s}^k,]^T$.

The modelling of the master was done using the assumed modes method. The slave manipulators are modelled using the lumped parameters method. Consider the dynamics of the slave manipulators as

$$\begin{cases} \dot{y}_1^k = y_2^k \\ \dot{y}_2^k = h^k(y^k, t) + c^k(y^k, t)u^k \end{cases} \tag{8.18}$$

where $y^k \epsilon R^4$. The synchronisation error between the master manipulator (8.1) and slave manipulators (8.18) is defined as

$$e_s^k = y_1^k - g_m \tag{8.19}$$

where

$$\begin{cases} g_m = [\phi_{1m}\ \phi_{2m}\ \xi_{1m}\ \xi_{2m}]^T \\ u_{1m} = w_{11}z_{11} + w_{12}z_{12} \\ u_{2m} = w_{21}z_{21} + w_{22}z_{22} \end{cases}$$

Now, our aim is to design an adaptive time-varying super-twisting global sliding mode control technique so that the synchronisation error converges to zero in the required time.

8.4.1 Projective synchronisation

The master manipulator is modelled using the assumed modes method while the non-identical manipulators are modelled using the lumped parameters method. Here, the projective synchronisation means that the states of the master manipulator are synchronised accordingly to the respective states of the slave manipulators. It is defined as

$$e_{is}^k = g_{is}^k - g_{im} \tag{8.20}$$

where $g_{im} = [\phi_{1m}, \phi_{2m}, \xi_{1m}, \xi_{1m}]^T$. Moreover, the elastic deflection of the master is projected to that of the slaves as

$$\begin{cases} \xi_{1s} \longrightarrow \xi_{1m} = w_{11m} z_{11m} + w_{12m} z_{12m} \\ \xi_{2s} \longrightarrow \xi_{1m} = w_{21m} z_{21m} + w_{22m} z_{22m} \end{cases} \tag{8.21}$$

The synchronisation of the master and the slave manipulator is achieved when (8.22) is satisfied

$$\lim_{t \to \infty} \|e_{is}^k(t)\| = 0 \tag{8.22}$$

8.4.2 Adaptive time-varying super-twisting global SMC

The initial step is to design normal sliding surface dynamics. The individual sliding surface is designed as

$$s_{is}^k = \dot{e}_{is}^k + c_{is}^k e_{is}^k \tag{8.23}$$

The next step is to design an equivalent sliding surface.

$$S_s^k = s_{1s}^k + a_{1s}^k s_{2s}^k + a_{2s}^k s_{3s}^k + a_{3s}^k s_{4s}^k \tag{8.24}$$

where c_{is}^k and a_{is}^k are the sliding surface gains.

Now, a global sliding surface is designed in (8.25)

$$\sigma_s^k(t) = G_s^k\{S_s^k(t) - e^{\Omega_s^k t} S_s^k(0)\} \tag{8.25}$$

where S_s^k is defined in (8.23) and G_s^k, Ω_s^k are the user defined positive gains.

In the global sliding mode, the sliding mode equation can be written as

$$G_s^k\{S_s^k(t) - e^{\Omega_s^k t} S_s^k(0)\} = 0 \tag{8.26}$$

since, the constants G_s^k are positive which satisfy the Hurwitz polynomial as in (8.8) and also satisfy the condition given in [20]. Thus, on the global sliding surface (8.24), the sliding mode dynamics (8.26) may converge to the origin. Thus, the convergence of error dynamics (8.20) can be shown by using (8.17), (8.24) and (8.25).

Next an appropriate GSMC based controller is designed to stabilise (8.25) at the origin. From (8.25)

$$\begin{cases} \dot{\sigma}_s^k(t) = G_s^k\{\dot{S}_s^k(t) + \Omega_s^k e^{\Omega_s^k t} S_s^k(0)\} \\ \quad = G_s^k\{\dot{s}_{1s}^k(t) + a_{1s}^k \dot{s}_{2s}^k(t) + a_{2s}^k \dot{s}_{3s}^k(t) + a_{3s}^k \dot{s}_{4s}^k(t) \\ \quad + \Omega_s^k e^{\Omega_s^k t}(\dot{s}_{1s}^k(0) + a_{1s}^k \dot{s}_{2s}^k(0) + a_{2s}^k \dot{s}_{3s}^k(0) \\ \quad + a_{3s}^k \dot{s}_{4s}^k(0))\} \\ \quad = [-b_1^k \dot{\phi}_{1s}^k + b_2^k \xi_{1s}^k - \dot{g}_m(1) - a_1^k b_4^k \dot{\phi}_{2s}^k + a_1^k b_5^k \xi_{2s}^k \\ \quad - \dot{g}_m(2) + a_2^k b_1^k \dot{\phi}_{1s}^k - a_2^k b_7^k \xi_{1s}^k - a_2^k w_{11m} \dot{g}_m(3) \\ \quad - a_2^k w_{12m} \dot{g}_m(4) + a_3^k b_4^k \dot{\phi}_{2s}^k - a_3^k b_8^k \xi_{2s}^k \\ \quad - a_3^k w_{21m} \dot{g}_m(5) - a_3^k w_{22m} \dot{g}_m(6) + c_{1s}^k \dot{e}_{1s}^k \\ \quad + a_1^k c_{2s}^k \dot{e}_{2s}^k + a_2^k c_{3s}^k \dot{e}_{3s}^k + a_3^k c_{4s}^k \dot{e}_{4s}^k \\ \quad + \Omega_s^k e^{\Omega_s^k t} S_s^k(0)] + [b_3^k + a_1^k b_6^k - a_2^k b_3^k - a_3^k b_6^k]u \end{cases} \tag{8.27}$$

where $\dot{g}_m$ is defined in (8.1). $\dot{g}_m(i)$ are the elements of $\dot{g}_m$. The necessary condition for the existence of error dynamics on the sliding surface (8.25) and for them to remain on it, it must satisfy $\dot{\sigma}_s^k = 0$. Now, the equivalent control input can be defined as

$$\begin{cases} u_{eq(s)}^k = (\frac{-1}{E_3})[-b_1^k \dot{\phi}_{1s}^k + b_2^k \xi_{1s}^k - \dot{g}_m(1) - a_1^k b_4^k \dot{\phi}_{2s}^k \\ \quad + a_1^k b_5^k \xi_{2s}^k - \dot{g}_m(2) + a_2^k b_1^k \dot{\phi}_{1s}^k - a_2^k b_7^k \xi_{1s}^k \\ \quad - a_2^k w_{11m} \dot{g}_m(3) - a_2^k w_{12m} \dot{g}_m(4) + a_3^k b_4^k \dot{\phi}_{2s}^k \\ \quad - a_3^k b_8^k \xi_{2s}^k - a_3^k w_{21m} \dot{g}_m(5) - a_3^k w_{22m} \dot{g}_m(6) \\ \quad + \hat{c}_{1s}^k \dot{e}_{1s}^k + a_1^k \hat{c}_{2s}^k \dot{e}_{2s}^k + a_2^k \hat{c}_{3s}^k \dot{e}_{3s}^k + a_3^k \hat{c}_{4s}^k \dot{e}_{4s}^k \\ \quad + \Omega_s^k e^{\Omega_s^k t} S_s^k(0)] \end{cases} \tag{8.28}$$

where $E_3 = b_3^k + a_1^k b_6^k - a_2^k b_3^k - a_3^k b_6^k$. In the equivalent control law (8.28), the parameters $\hat{c}_{is}^k$ of the sliding surfaces are the estimates of c_{is}^k. These are updated with the following adaptation laws:

$$\begin{cases} \dot{\hat{c}}_1^k = \sigma_s^k \dot{e}_{1s}^k \\ \dot{\hat{c}}_2^k = \sigma_s^k a_1^k \dot{e}_{2s}^k \\ \dot{\hat{c}}_3^k = \sigma_s^k a_2^k \dot{e}_{3s}^k \\ \dot{\hat{c}}_4^k = \sigma_s^k a_3^k \dot{e}_{4s}^k \end{cases} \tag{8.29}$$

The corrective control law is defined as

$$u_c^k = -\hat{\rho} sat(\sigma_s^k) + \Upsilon \sigma_s^k + \Theta |\sigma_s^k|^r \tag{8.30}$$

Θ and r are the positive constants. In (8.30), the corrective control gain $\hat{\rho}$ is the estimate of ρ which defines the boundedness $|P(g, \dot{g}, \ddot{g})| \leq \rho$. Since, the bounds of the system uncertainty change with respect to time, hence ρ is considered as unknown or updated with respect to time. The adaptation law is defined as follows [20]

$$\dot{\tilde{\rho}} = \dot{\hat{\rho}} = K_p^{-1}|\sigma_s^k| \tag{8.31}$$

where K_p is the adaptation gain and $\tilde{\rho} = \hat{\rho} - \rho$.

Using (8.29) and (8.30), control inputs can be defined as

$$u_s^k = u_{eq(s)}^k + u_c^k \tag{8.32}$$

Theorem 8.3. *The trajectory of the LPM modelled slave manipulator follows the AMM modelled controlled master manipulator and remains on the sliding surface $\dot{\sigma}_s^k = 0$. The slave manipulators (8.17) synchronise with the controlled master manipulator (8.1) using the sliding surfaces (8.23), (8.24), (8.25) and synchronisation controller (8.32) along with the adaptation law defined by (8.29).*

Proof. Considering the Lyapunov function candidate

$$V_3^k(\sigma_s^k, \tilde{c}_{is}^k, \tilde{\rho}) = \frac{1}{2}(\sigma_s^k)^2 + \frac{1}{2}\sum_{i=1}^{4} \tilde{c}_{is}^k + \frac{1}{2}k_p\tilde{\rho}^2 \tag{8.33}$$

and using (8.27) with (8.29), (8.31) with (8.32) we get,

$$\begin{cases} \dot{V} = -k_1(\tilde{c}_{1s}^k)^2 - k_2(\tilde{c}_{2s}^k)^2 - k_3(\tilde{c}_{3s}^k)^2 - k_4(\tilde{c}_{4s}^k)^2 \\ \quad - \sigma_s^k(\hat{\rho} sat(\sigma_s^k) + \Upsilon\sigma_s^k + \Theta|\sigma_s^k|^r) - \sigma_s^k P(g, \dot{g}, \ddot{g}) \\ \leq -\Upsilon(\sigma_s^k)^2 - \Theta|\sigma_s^k|^{r+1} + |\sigma_s^k|P(g, \dot{g}, \ddot{g}) - \rho|\sigma_s^k| + Y \\ = Y - \Upsilon(\sigma_s^k)^2 - \Theta|\sigma_s^k|^{r+1} - \{\rho - |P(g, \dot{g}, \ddot{g})|\}|\sigma_s^k| \\ = Y - \Upsilon(\sigma_s^k)^2 - \Theta|\sigma_s^k|^{r+1} \end{cases} \tag{8.34}$$

where $Y = -k_1(\tilde{c}_{1s}^k)^2 - k_2(\tilde{c}_{2s}^k)^2 - k_3(\tilde{c}_{3s}^k)^2 - k_4(\tilde{c}_{4s}^k)^2$, $|P| \leq \rho$.

Hence, the trajectory of the master manipulator (8.1) follows the slave manipulator (8.17) within a certain interval of time. □

Remark 8.4. Although to avoid the chattering, the saturation function is used in (8.32) of the corrective control law it may cause some robustness and convergence issues. Here, super-twisting control algorithm is used to avoid the chattering problem and robustness issue. Thus, the corrective control input is modified as

$$\begin{cases} u_c^k = \aleph + \Upsilon\sigma_s^k + \Theta|\sigma_s^k|^r + \nu|\sigma_s^k|^n sgn(\sigma_s^k) \\ \quad \dot{\aleph} = \hat{\rho} sgn(\sigma_s^k) \end{cases} \tag{8.35}$$

where $\nu > 0, n = \frac{1}{2} > 0$.

Table 8.1: Initial conditions used with different equations.

Equation No.	Initial Conditions
(8.1)	$g_m = (0.5, 0.5, 0, 0, 0, 0)^T$
(8.1)	$\dot{g}_m = (0, 0, 0, 0, 0, 0)^T$
(8.17)	$g_s^1 = (0.1, 0.1, 0, 0)^T$
(8.17)	$\dot{g}_s^1(0) = (0, 0, 0, 0)^T$
(8.17)	$g_s^2 = (0.01, 0.01, 0, 0)^T$
(8.17)	$\dot{g}_s^2(0) = (0, 0, 0, 0)^T$
(8.17)	$g_s^3 = (1X10^{-4}, 1X10^{-4}, 0, 0)^T$
(8.17)	$\dot{g}_s^3(0) = (0, 0, 0, 0)^T$
(8.29)	$c_{1s}^k = c_{2s}^k = c_{3s}^k = c_{4s}^k = 0$
(8.35)	$N(0) = 0$

8.5 Results and discussion

Projective synchronisation has been achieved between the 12 states of the master manipulator and the 8 states of the slave manipulators. The elastic deflection of each link of the master manipulator is captured by 4 states while the same is captured by 2 states for the slave manipulators. Thus, 8 states of the master manipulator will project to 4 states of the slave manipulator. The remaining 4 states of the master will project to the remaining 4 states of the slave. This will result in the projective synchronisation between the 12 states of the master and the 8 states of the slaves.

The performances of the objective are tested in two steps. The master manipulator is required to track a desired/reference signal first. Then, projective synchronisation is achieved between the controlled master and the slave manipulators. The desired/reference signal considered here is given in (8.36)

$$\phi_{id} = \frac{\pi}{6} - \frac{7}{6}e^{-1.5t} + \frac{7}{11}e^{-2t} \tag{8.36}$$

We have assumed a 0.145 kg payload during simulations of all the cases given below. The parameters of the master, slave-1, slave-2 and slave-3 manipulators are given in Tables 3.1 and 6.2, respectively. The simulations are carried out in a MATLAB® 14a environment using ode45 solver with a step size of 0.02. The initial conditions of the master and slave manipulators are given in Table 8.1. The values of the parameters of the slave manipulators are, $b_1^{1,2} = 75.6721$, $b_2^{1,2} = 3.2358$, $b_3^{1,2} = 4.3867$, $b_4^{1,2} = 1026.12$, $b_5^{1,2} = 387.6975$, $b_6^{1,2} = 59.4266$, $b_7^{1,2} = 274.2314$, $b_8^{1,2} = 3883.41$, $b_1^3 = 20.7965$, $b_2^3 = 7.9627$, $b_3^3 = 1.2953$, $b_4^3 = 22.3788$, $b_5^3 = 4.0997$, $b_6^3 = 1.3939$, $b_7^3 = 412.2217$, $b_8^3 = 267.3053$, where superscript denotes the number of the slave.

8.5.1 Results on trajectory tracking of the master manipulator

A GSMC is used to control the master manipulator to track the desired trajectory. The constant gains of the sliding surfaces and the controller gains for controlling the master manipulator are given as $c_{1m} = 5$, $c_{2m} = 15$, $\rho_{1m} = 5$ and $\rho_{2m} = 15$. χ_i and μ_i are the positive constants. Here, we have taken $\chi_i = 10$, $\mu_i = 1$. Figure 8.2 shows the tracking error between the desired trajectory (8.36) and the actual trajectory of the master manipulator (8.1). The tracking errors have reached zero at 0.9 s,

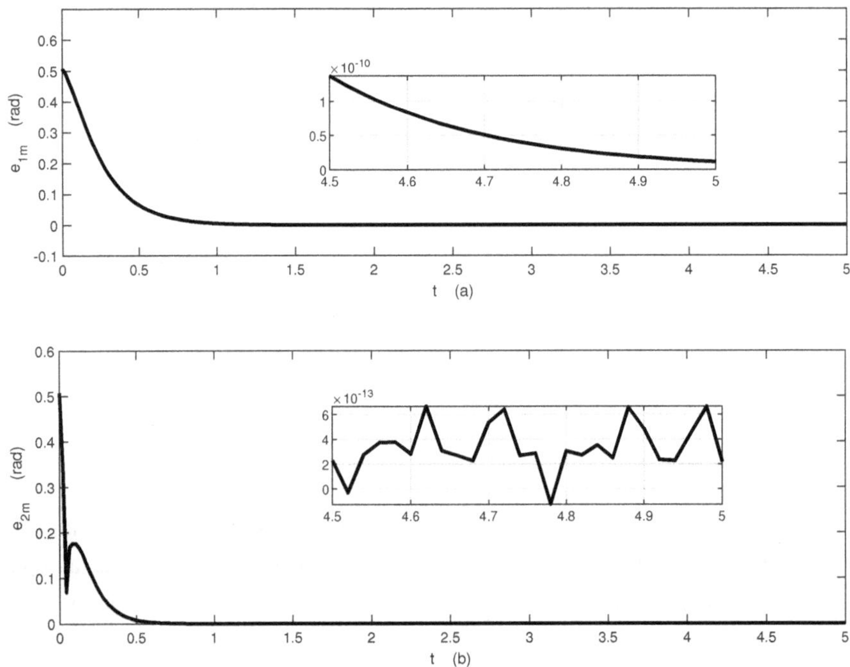

Figure 8.2: Trajectory tracking errors between the desired trajectory (8.36) and the master manipulator for a) link-1 and b) link-2.

0.5 s for link-1 and link-2, respectively. The tracking errors are of the order of 10^{-10} and 10^{-12} at their steady state. The tip deflections of link-1 and link-2 of the master manipulator are shown in Fig. 8.3. The maximum tip deflections of links 1 and 2 are $18\text{x}10^{-3}$ mm at 0.7 s and $4.14\text{x}10^{-5}$ mm at 0.3 s, respectively. From Figs. 8.2 and 8.3, it is noted that the master manipulator tracks the desired trajectory successfully along with the suppression of tip deflection. As the master manipulator tracks the desired trajectory using a GSMC, the designed sliding surfaces for the normal SMC and GSMC are shown in Figs. 8.4 and 8.5, respectively. The reaching time of the sliding surfaces of the normal SMC for links 1 and 2 are 0.5 s and 0.4 s, respectively. On the other hand, the reaching time of the sliding surfaces of the GSMC is almost eliminated, as shown in Fig. 8.5. However, the reaching time is slightly present as it depends on the values of χ and μ. As χ and μ increases, the reaching time is reduced but this hampers the deflections of the manipulator. The manipulator elastic deflections tend to last longer. Hence, we have taken the value of χ and μ in such a way that the deflections are suppressed properly. The control torques of links 1 and 2 are shown in Fig. 8.6.

8.5.2 Results on projective synchronisation

This subsection describes the projective synchronisation between the controlled master and the three slave manipulators. An adaptive time-varying super-twisting GSMC (A-TV-ST-GSMC) is used for projective synchronisation between the controlled master and the slave manipulators. The gains of the sliding surfaces and the corrective control laws are presumed to be time-varying and are adapted in the proposed A-TV-ST-GSMC algorithm. Three different cases of projective synchronisation

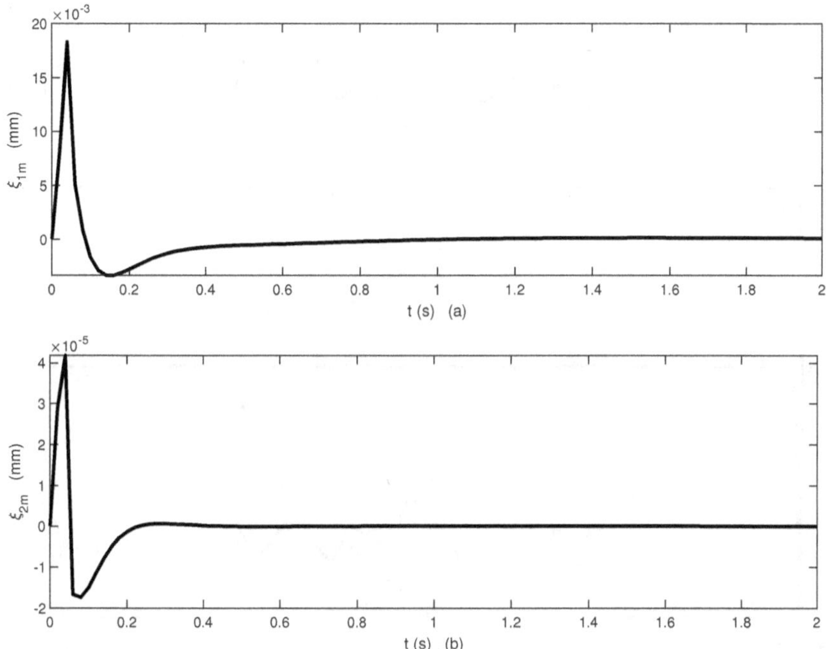

Figure 8.3: Tip deflections of the master manipulator for a) link-1 and b) link-2.

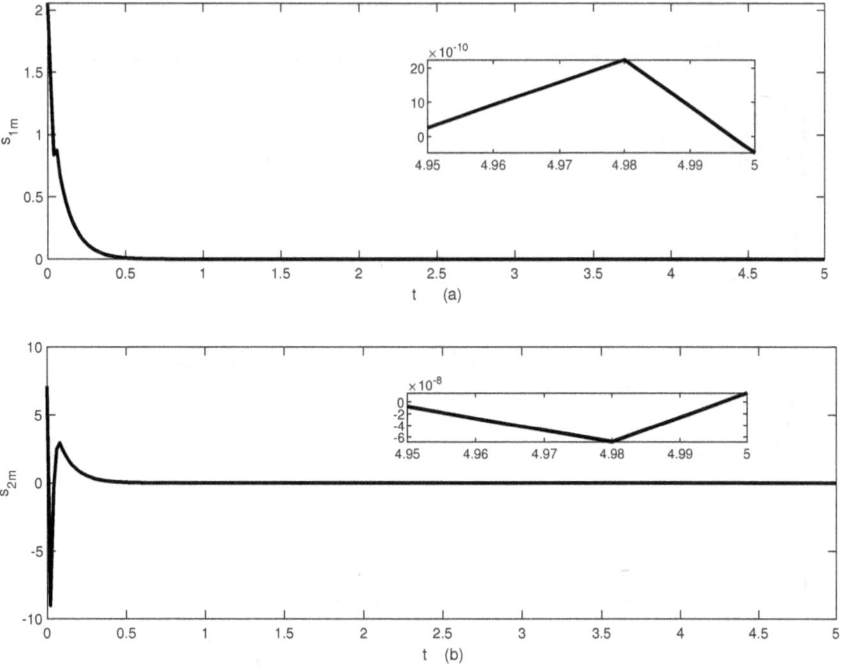

Figure 8.4: Sliding surfaces of the master manipulator for a) link-1 and b) link-2.

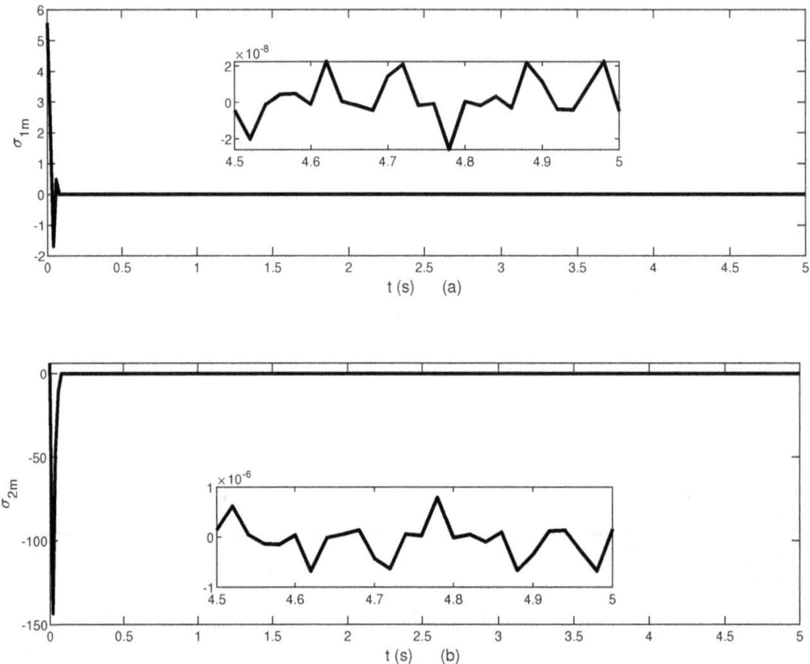

Figure 8.5: Global sliding surfaces of the master manipulator for a) link-1 and b) link-2.

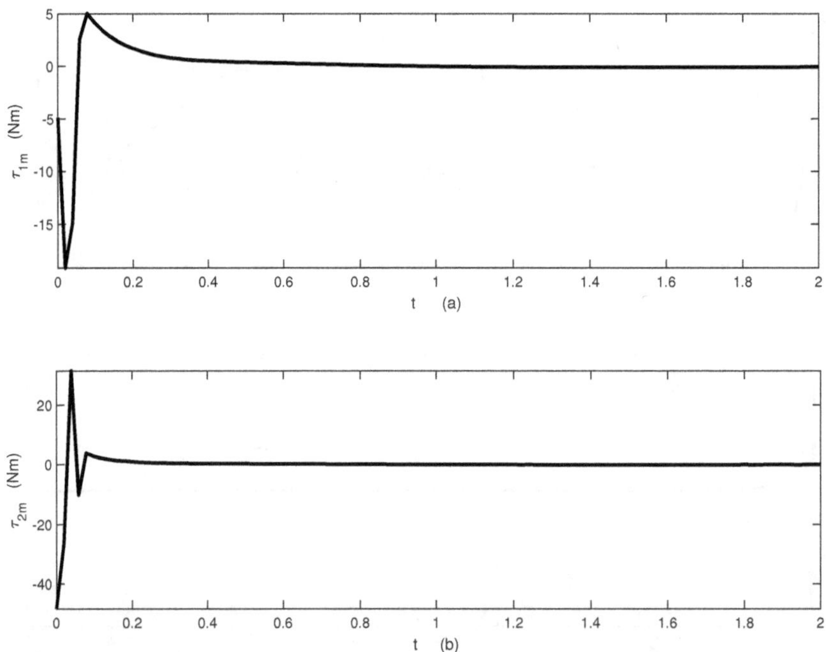

Figure 8.6: Control torques for tracking the desired trajectory by the master manipulator for a) link-1 and b) link-2.

are considered. The first and the second cases are between the controlled master and the identical slaves and the third case is between the controlled master and the non-identical slave manipulators. The results are shown where the states of the controlled master manipulator are projected to the states of the slave manipulators. In the simulation results, e_{is}^{k} denotes the synchronisation errors of the ith link for the k slaves. The projective synchronisation trajectory tracking errors between the controlled master and the slave manipulators are shown in Fig. 8.7. The projective synchronisation tip deflection errors between the master and the slaves of link-1 are shown in Fig. 8.8. One state of the slaves is projected to two states of the controlled master. Figure 8.9 shows the projective tip deflection suppression of link-2. The sliding surfaces are shown in Figs. 8.10, 8.11 and 8.12. Figure 8.13 shows the torque input used for the synchronisation of the manipulators. The adaptation of the gains of the sliding surfaces and the switching control laws are shown in Figs. 8.14, 8.15 and 8.16.

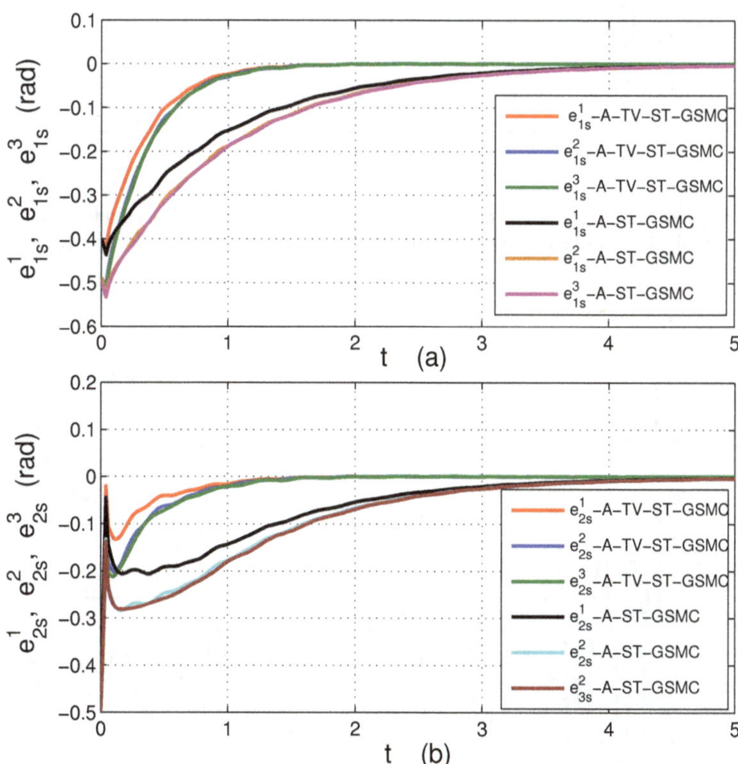

Figure 8.7: Comparison of projective synchronisation errors between the controlled master and slave manipulators for the proposed controller and controller in [20] for link-1 and link-2.

8.5.2.1 Comparison of the proposed controller with the controller of [20]

To the best of our knowledge, we have not found any paper which deals with the projective synchronisation of AMM modeled TLFM and LPM modeled TLFM in the literature. The proposed projective synchronisation strategy is achieved by using an A-TV-ST-GSMC technique. Hence, to

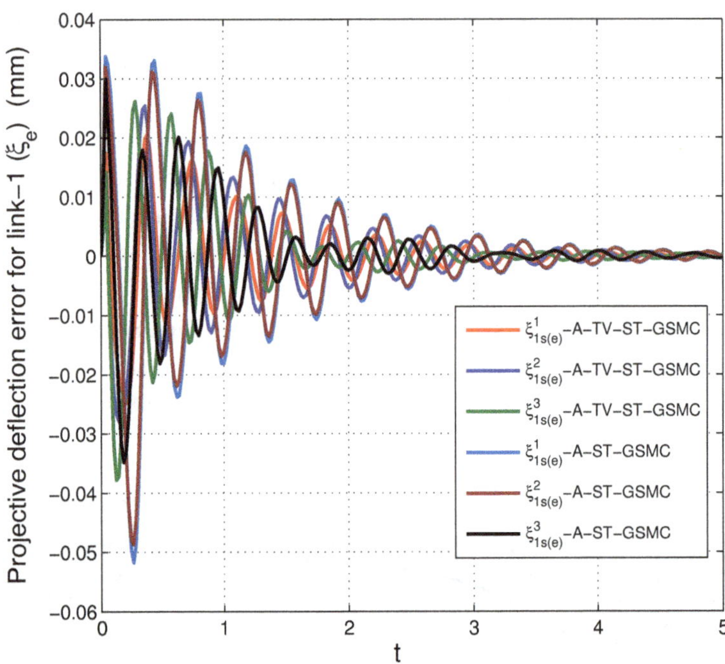

Figure 8.8: Comparison of projective synchronisation deflection errors between the controlled master and slave manipulators using the proposed controller and the controller in [20] for link-1.

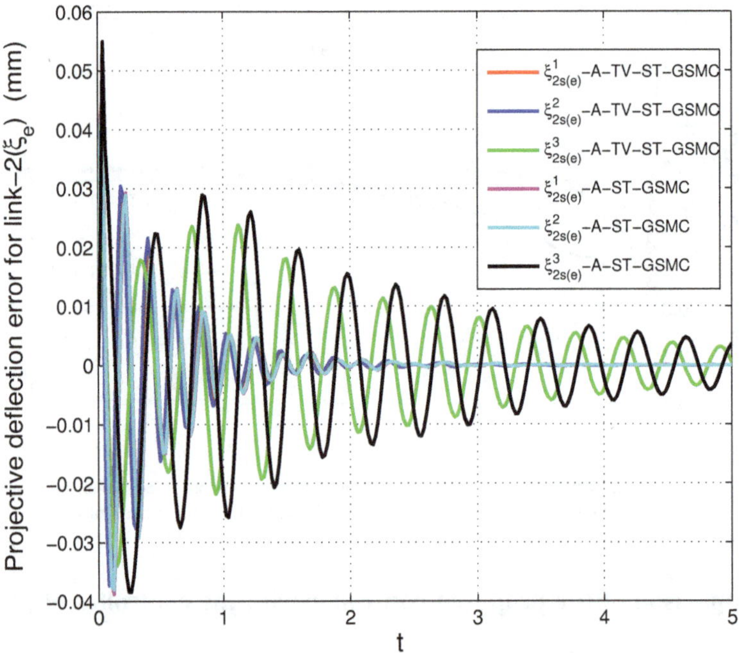

Figure 8.9: Comparison of projective synchronisation deflection errors between the controlled master and slave manipulators using the proposed controller and the controller in [20] for link-2.

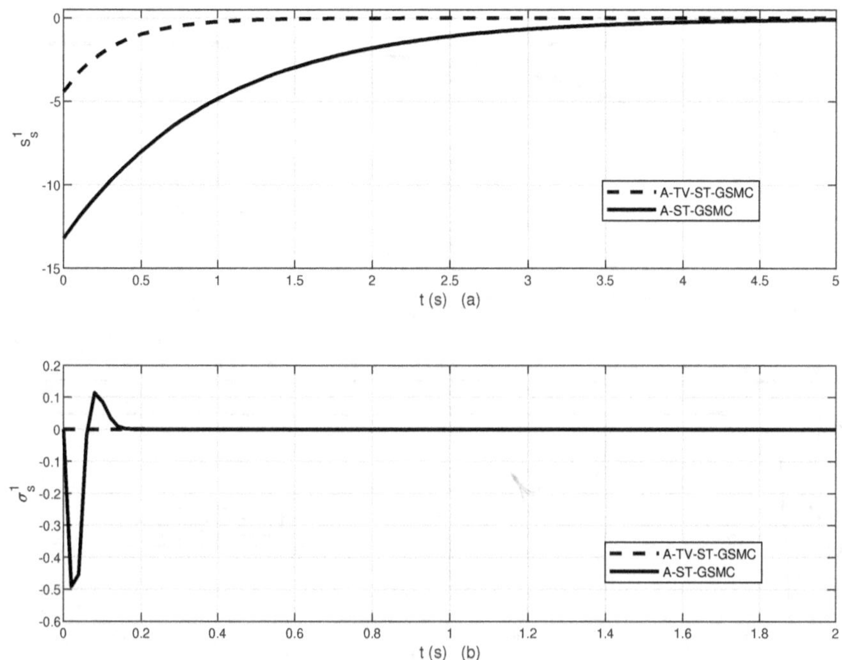

Figure 8.10: Comparison of the equivalent and the global sliding surfaces for projective synchronisation between the controlled master and slave-1 manipulator using the proposed controller and the controller in [20].

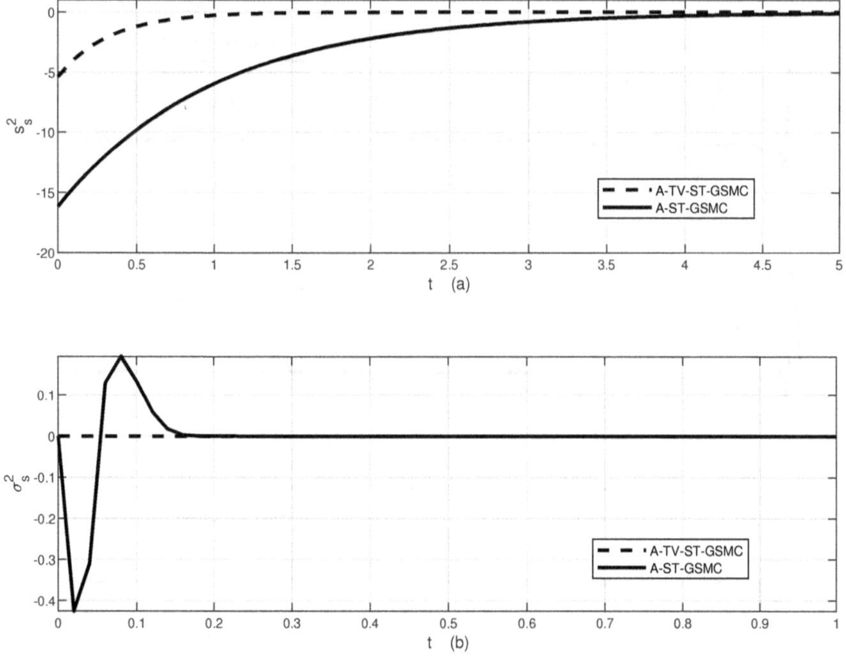

Figure 8.11: Comparison of the equivalent and the global sliding surfaces for projective synchronisation between the controlled master and the slave-2 manipulator using the proposed controller and the controller in [20].

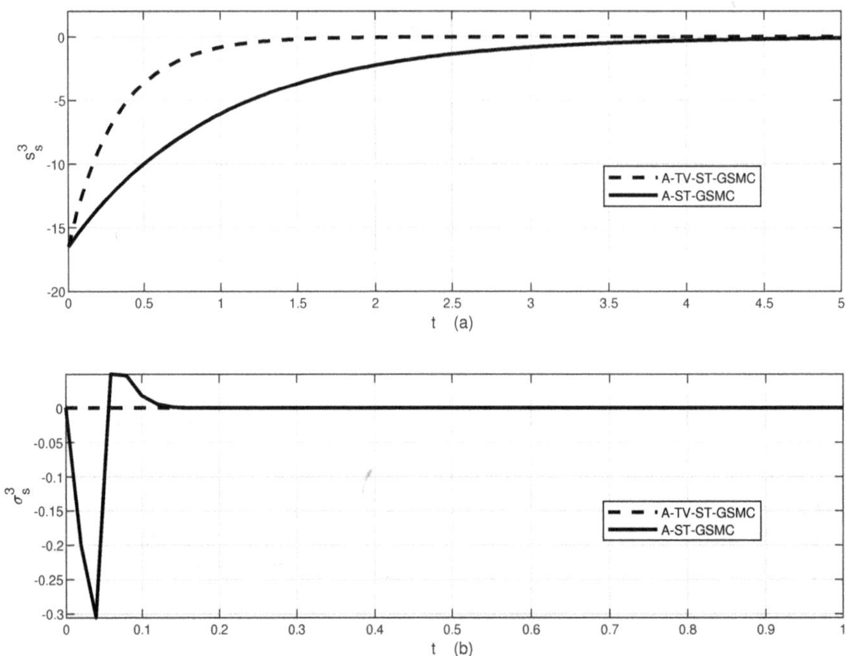

Figure 8.12: Comparison of the equivalent and the global sliding surfaces for projective synchronisation between the controlled master and slave-3 manipulator using the proposed controller and the controller in [20].

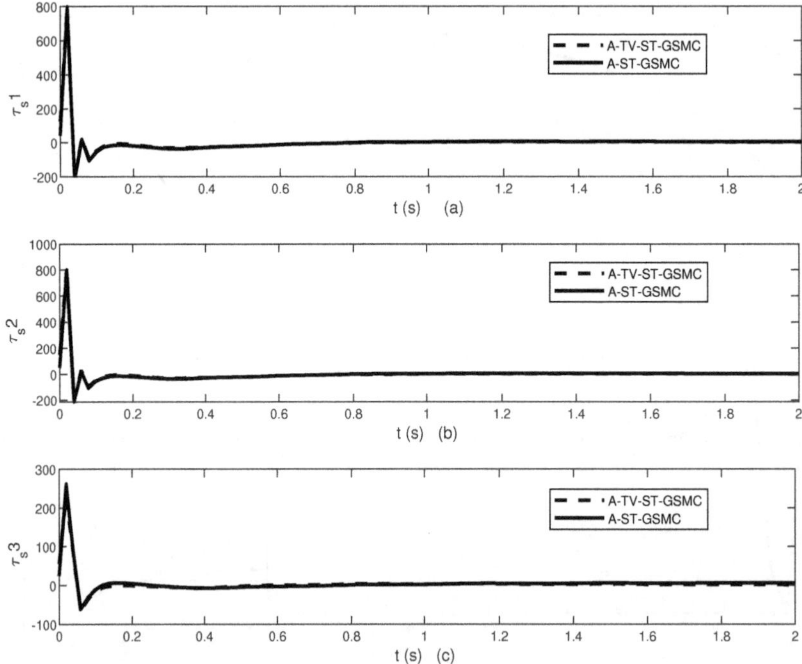

Figure 8.13: Comparison of the torque for projective synchronisation using the proposed controller and the controller in [20].

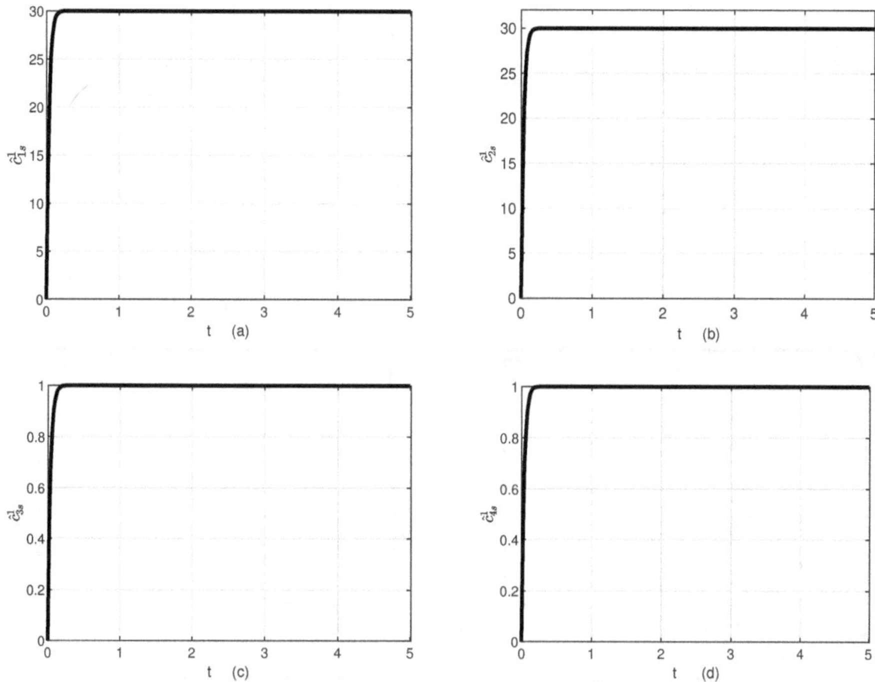

Figure 8.14: Time responses of the estimated parameters of slave-1.

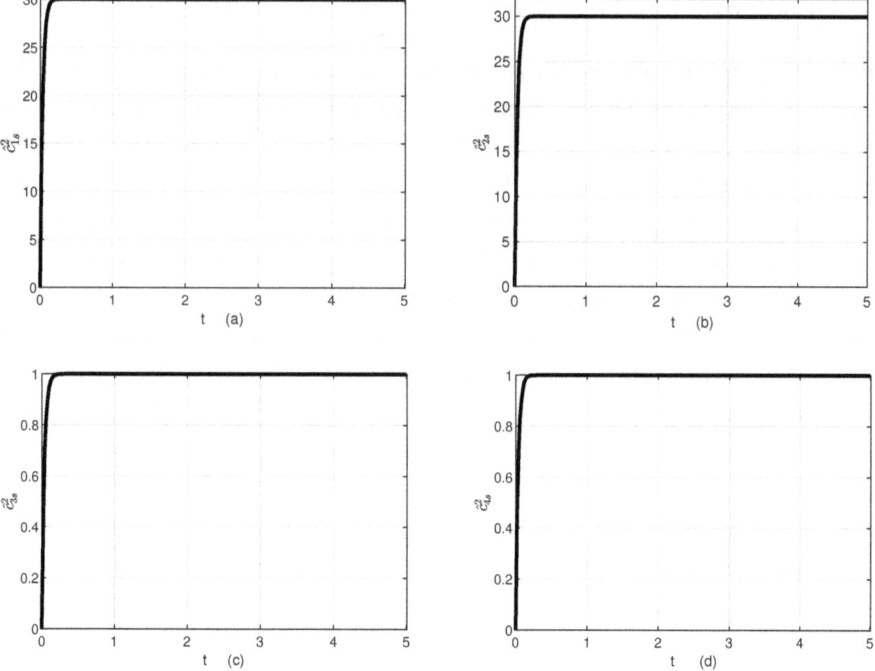

Figure 8.15: Time responses of the estimated parameters of slave-2.

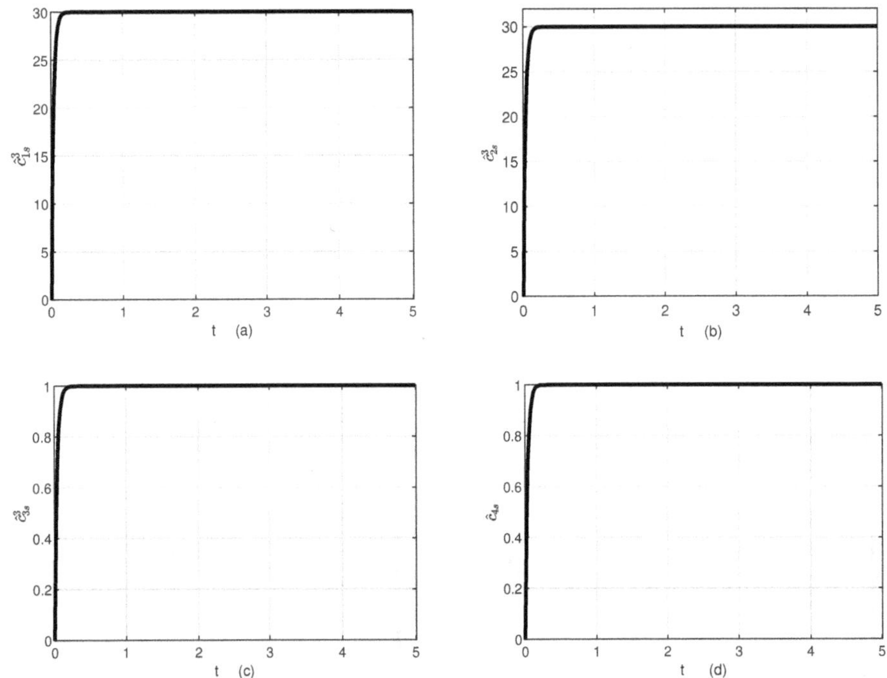

Figure 8.16: Time responses of the estimated parameters of slave-3.

compare the performance of the proposed control technique, a recently proposed adaptive super-twisting global SMC (A-ST-GSMC) controller by Mobayen and Tchier (2016) [20] is considered.

It is seen from Fig. 8.7 that the projective synchronisation errors settling time of link-1 for all the three cases by using the proposed controller is $1.8\ s$ whereas, in the case of the controller of [20], it is $4\ s$. For link-2, the synchronisation is achieved at $1.5\ s$ using our controller but the controller of [20] takes $4.2\ s$. It is apparent from the Fig. 8.8 that the tip deflection amplitude of link-1 of the three slaves using the proposed controller reaches its maximum in the range of -0.038 to $0.032\ mm$ and the same is the range of -0.052 to $0.034\ mm$ for the controller in [20]. The tip deflection suppression in link-2 using both the controllers is almost similar as shown in Fig. 8.9. It is evident from Figs. 8.10, 8.11 and 8.12 that the sliding surface reaching time in case of the proposed controller is much lower than that of the compared controller. The torque input requirements of both the proposed controller and the controller of [20] are shown in Fig. 8.13. However, as observed from Fig. 8.13, the variation in our controller is less. It is seen from Figs. 8.14, 8.15 and 8.16 that the constant gain requirement using the proposed controller is less as compared with the controller of [20]. From all these figures, it is clear that the proposed controller has a lower synchronisation time, lesser tip deflection and smother control input when compared with the performances of the controller in [20]. The comparison of the performances of the controllers are given in Table 8.2.

Table 8.2: Comparison on performances of the controllers.

	Performances of the controllers					
Control techniques	Controller in [20]			Proposed controller		
Performance Indices	Slave-1	Slave-2	Slave-3	Slave-1	Slave-2	Slave-3
Control energy (2-norm)	877.262	880.714	314.127	**858.961**	**864.737**	**262.658**
Total Variance (TV)	2222.9	2228.32	654.317	2222.52	2222.9	**583.761**
IAE of tracking error (link-1)	0.415	0.5106	0.213	**0.151**	**0.188**	**0.191**
IAE of tracking error (link-2)	0.331	0.4232	0.107	**0.0675**	**0.1**	**0.106**
IAE of link deflection (link-1)	0.0349	0.0326	0.0206	**0.0215**	**0.0206**	**0.0205**
IAE of link deflection (link-2)	0.0151	**0.0149**	0.0407	**0.0144**	0.0152	**0.0404**

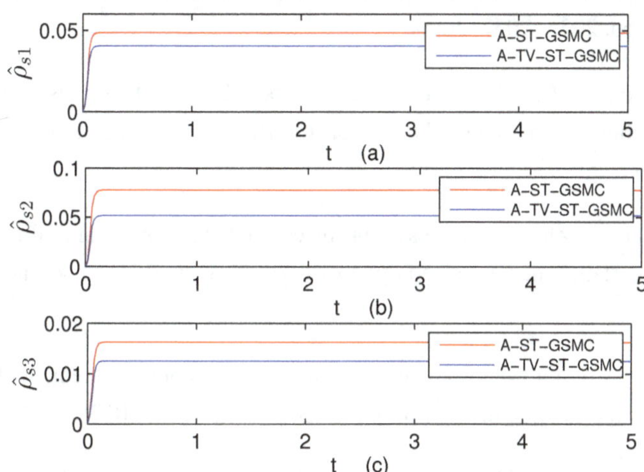

Figure 8.17: Comparison of the time responses of the estimated ρ using the proposed controller and the controller in [20] for slave-1, slave-2 and slave-3.

8.6 Chapter summary

In this chapter, three cases of projective synchronisation between the controlled master and slave TLFMs are presented in this paper. An identical master and slave combination is considered in the first two cases. However, the slave is presumed to be different from the master in the last case. The objective is achieved in two steps. First a global SMC is designed for tracking control of a master two-link flexible manipulator. Next, a new adaptive time-varying super-twisting global SMC is proposed for the synchronisation between a slave manipulator with the controlled master manipulator. An AMM method is used for modelling the master manipulator and the LPM method is used for modelling the slave manipulator. Therefore, a projective synchronisation strategy is designed for the slave and master manipulators having eight and twelve states, respectively. Gains of the corrective control laws of the SMC are estimated online and a super twisting algorithm is designed to reduce the chattering problem. A global SMC based approach is used for the synchronisation to reduce the sliding surface reaching time. This helps to increase the robustness of the controller. The performance of the proposed controller for the synchronisation is compared with a controller reported in [20]. It is seen that the proposed scheme performs better in terms of synchronisation error, reaching time and control energy in comparison with the controller of [20].

References

[1] Y. Yang, C. Hua, J. Li, and X. Guan. Fixed-time Coordination Control for Bilateral Telerobotics System with Asymmetric Time-varying Delays. *J. Intell. Robot. Syst.*, vol. 86, pp. 447–466, 2016.

[2] F. Hashemzadeh, I. Hassanzadeh, M. Tavakoli, and G. Alizadeh. Adaptive control for state synchronization of nonlinear haptic telerobotic systems with asymmetric varying time delays. *J. Intell. Robot. Syst. Theory Appl.*, vol. 68, no. 3, pp. 245–259, 2012.

[3] P. F. Hokayem and M. W. Spong. Bilateral teleoperation: An historical survey. *Automatica*, vol. 42, no. 12, pp. 2035–2057, 2006.

[4] D. Zhao, S. Li, and Q. Zhu. A new TSMC prototype robust nonlinear task space control of a 6 DOF parallel robotic manipulator. *Int. J. Control. Autom. Syst.*, vol. 8, no. 6, pp. 1189–1197, 2010.

[5] D. Zhao, S. Li, and Q. Zhu. Adaptive synchronised tracking control for multiple robotic manipulators with uncertain kinematics and dynamics. *Int. J. Syst. Sci.*, vol. 47, no. 4, pp. 791–804, 2015.

[6] H. Wang. Passivity based synchronization for networked robotic systems with uncertain kinematics and dynamics. *Automatica*, vol. 49, no. 3, pp. 755–761, 2013.

[7] A. Rodriguez-Angeles and H. Nijmeijer. Mutual synchronization of robots via estimated state feedback: A cooperative approach. *IEEE Trans. Control Syst. Technol.*, vol. 12, no. 4, pp. 542–54, 2016.

[8] Y. Su, D. Sun, L. Ren, and J. K. Mills. Integration of saturated PI synchronous control and PD feedback for control of parallel manipulators. *IEEE Trans. Robot.*, vol. 22, no. 1, pp. 202–207, 2006.

[9] W. Shang, S. Cong, Y. Zhang, and Y. Liang. Active Joint Synchronization Control for a 2-DOF Redundantly Actuated Parallel Manipulator. *IEEE Trans. Control Syst. Technol.*, vol. 17, no. 2, pp. 416–423, 2009.

[10] M. Khazaee, A. H. D. Markazi, and E. Omidi. Adaptive Fuzzy Predictive Sliding Control of Uncertain Nonlinear systems with Bound-known Input Delay. *ISA Transaction*, vol. 59, pp. 314–324, 2015.

[11] Q. Cao, S. Li, D. Zhao, and Z. Wang. Finite-Time Motion/Force Control for Motion Synchronization of Multiple Manipulators. *In Proceedings of the 33rd Chinese Control Conference, Nanjing, China*, pp. 2121–2126, 2014.

[12] D. Zhao, S. Li, and F. Gao. Finite time position synchronised control for parallel manipulators using fast terminal sliding mode. *Int. J. Syst. Sci.*, vol. 40, no. 8, pp. 829–843, 2009.

[13] L. Li, L. Sun, S. Zhang, and Q. Yang. Speed tracking and synchronization of multiple motors using ring coupling control and adaptive sliding mode control. *ISA Trans.*, vol. 58, pp. 635–649, 2015.

[14] D. Zao, S. Li, F. Gao, and Q. Zhu. Robust adaptive terminal sliding mode-based synchronised position control for multiple motion axes systems. *IET Control Theory Appl.*, vol. 3, no. 3, pp. 136–150, 2009.

[15] D. Zhao, T. Zou, S. Li, and Q. Zhu. Adaptive backstepping sliding mode control for leader-follower multi-agent systems. *IET Control Theory Appl.*, vol. 8, no. 6, pp. 1109–1117, 2012.

[16] Z. Q. Zhao D, Li C. Low-pass-filter-based position synchronization sliding mode control for multiple robotic manipulator systems. *Proc. Inst. Mech. Eng. Part I J. Syst. Control Eng*, pp. 1136–1148, 2011.

[17] Z. Q. Zhao D. Position synchronised control of multiple robotic manipulators based on integral sliding mode. *Int. J. Syst. Sci.*, vol. 45, pp. 556–570, 2014.

[18] C. Weng, and W. Yu. H_∞ tracking adaptive fuzzy integral sliding mode control for parallel manipulators. *Fuzzy Sets Syst.*, vol. 248, pp. 1–38, 2014.

[19] L. Liu, Z. Han, and W. Li. Global sliding mode control and application in chaotic systems. *Nonlinear Dyn.*, vol. 56, no. 2, pp. 193–198, 2008.

[20] S. Mobayen, and F. Tchier. Design of an adaptive chattering avoidance global sliding mode tracker for uncertain non-linear time-varying systems. *Trans. Inst. Meas. Control*, pp. 1–12, 2016.

[21] S. Mobayen. A Novel Global Sliding Mode Control Based on Exponential Reaching Law for a Class of Underactuated Systems With External Disturbances. *J. Comput. Nonlinear Dyn.*, vol. 11, no. 2, pp. 1–9, 2015.

Chapter 9

Conclusions and Future Work

There has been a keen interest among the researchers in the domain of higher order sliding mode control and synchronisation of flexible manipulators. In spite of all the advances from the perspective of the control, the issue will remain open for research in the coming years. This book focuses on the development of new sliding mode control approaches and the synchronisation of two-link flexible manipulators with varying payloads and parameter disturbances. The designed control techniques are implemented on the developed mathematical model of a two-link flexible manipulator. This chapter concludes the book and some future scopes for the extension of the work described in this book are also highlighted.

9.1 Summary of the book work

This book mainly investigates the development of new sliding mode control techniques to control the tip trajectory tracking while suppressing its tip deflections along with varying payloads. The robustness of the designed controllers is also checked. The singular perturbation is explored with the assumed modes method modelling for decomposing the system into a slow and fast dynamics. The controllers are also designed for the synchronisation of two-link flexible manipulators. The synchronisation between the controlled master and the slave manipulator(s) is achieved as (i) lumped parameter-lumped parameter modelling, (ii) assumed modes-assumed modes modelling and (iii) assumed modes-lumped parameter modelling of two-link flexible manipulators. Various control techniques such as (i) conventional sliding mode control, (ii) modified adaptive sliding mode control, (iii) adaptive sliding mode control, (iv) linear matrix inequality based sliding mode control, (v) linear matrix inequality based state feedback controller, (vi) backstepping controller, (vii) second order proportional integral derivative terminal sliding mode control, (viii) global sliding mode control and (ix) adaptive time-varying super-twisting global sliding mode control are used to achieve the above-mentioned trajectory tracking and synchronisation. A summary of the book is presented here:

- Introduction to the two-link flexible manipulator is described in Chapter 1.
- A detailed literature survey is done for two-link flexible manipulators where the research gaps are found and motivation of the book is formulated.
- The dynamic model of a two-link flexible manipulator is derived using Euler-Lagrange method in combination with assumed modes method. The mathematical model is validated in the free and forced conditions. From the simulation results, it is confirmed that the derived model is appropriate to represent the dynamics of the two-link flexible manipulator.
- A new second order sliding mode control is developed to control the tip trajectory tracking while simultaneously suppressing the tip deflection. This is done while subjecting to varying payloads. The second order sliding surfaces are designed in terms of the conventional sliding surfaces. The performances of the second order sliding mode control are compared with the conventional sliding mode control for different payloads of nominal as well as 0.3 kg. It is shown that the chattering in sliding surfaces and control inputs are removed. Fast tip trajectory tracking, quick tip deflection suppression and chattering free control inputs are considered as the performance metrics of the designed second order-SMC and is found to be better in comparison with the conventional SMC.
- The singular perturbation is explored with the assumed modes method for the dynamic modelling of a two-link flexible manipulator. An LMI based SMC and LMI based state feedback controller are designed of the slow and fast subsystems. The slow subsystem tracks the rigid dynamics and the fast subsystem tracks the fast dynamics. Another composite controller includes the design of adaptive SMC for the slow subsystem and backstepping for the fast subsystem with proper boundary conditions. The trajectory tracking and tip deflecting suppression along with the varying payloads are compared with an existing controller in the literature. The desired trajectory used is a chaotic signal for the second composite controller. The proposed second composite controller has smaller steady state errors, quick and smaller tip deflections and required lesser control efforts when compared with the controller in the literature.
- A generalised projective synchronisation technique is designed for the synchronisation of the controlled master and slave manipulators. The complete as well as partial scaling are done between the master-slave synchronisation for the exponentially varying as well as chaotic desired signal. A modified adaptive equivalent sliding mode control is designed for the synchronisation of the lumped parameter modelled master and slaves. The proposed controller is compared with three other variants of sliding mode controllers. The results indicate an excellent tracking accuracy as compared with other controllers while subjecting to varying payloads.

- A new second order PID terminal sliding mode control is designed for the synchronisation between the controlled master and slave manipulators. The synchronisation is done between assumed modes modelled two-link flexible manipulators. The robustness of the designed controller is checked with varying parameters from $\pm 5\%$ to $\pm 30\%$ uncertainties subjected to changes in payloads. The results show that the proposed control technique has the ability for a finite time convergence of synchronisation errors along with fast suppression of tip deflection and global stability of the manipulator dynamics.

- A projective synchronisation is done between a controlled master and three slave manipulators. The master is modelled by the assumed modes method and the slaves are modellled by the lumped parameters method. The twelve states controlled master is projected to eight states slaves. An adaptive time-varying super-twisting global sliding mode is designed for the synchronisation technique. The designed controller is compared with an existing controller. It is seen that the proposed scheme performs better in terms of synchronisation error, reaching time and control energy consumption in comparison with the controller in the literature.

9.2 Book contributions

- Development of the mathematical model and its validation in free and forced conditions.

- Development of the conventional and second order sliding mode control and to verify the control performances on achieving tip trajectory tracking along with the tip deflection suppression subjected to variable payloads.

- Development of dynamics segregation and designing of two composite controllers for the slow and fast subsystems. A LMI-SMC and LMI-SFC for the first composite controller and adaptive SMC and backstepping controller for the second composite controller while subjecting to varying payloads.

- Development of a modified adaptive equivalent sliding mode control for the generalised projective synchronisation with complete as well as half scaling considering exponentially time varying and chaotic trajectory signals subjected to varying payloads.

- Development of a new second order PID terminal sliding mode control for synchronisation between the assumed modes modelled TLFMs with change in payload and varying parameters uncertainty.

- Development of a new adaptive time-varying super-twisting global sliding mode control for projective synchronisation between the assumed modes modelled master and lumped parameter modelled slave manipulators.

9.3 Suggestions for future work

The proposed research work can be extended as follows in the near future:

- Experimental validation of the proposed controllers.
- Network controlled synchronisation scheme for a TLFM.
- Time delay control design of a synchronised network.

Index

U

V

For Product Safety Concerns and Information please contact our EU
representative GPSR@taylorandfrancis.com
Taylor & Francis Verlag GmbH, Kaufingerstraße 24, 80331 München, Germany

www.ingramcontent.com/pod-product-compliance
Lightning Source LLC
LaVergne TN
LVHW081317110826
845149LV00006B/1527

* 9 7 8 1 0 3 2 3 8 4 7 8 8 *